The Fumars Hot Spring, Azores, Portugal

Photo credit: Tim Ravenna

THE
TROUBLED WATERS
OF EVOLUTION

by

HENRY M. MORRIS, Ph.D.

President, Institute for Creation Research

C·L·P
PUBLISHERS
San Diego, California

The Troubled Waters of Evolution

First Edition Copyright © 1975 Henry M. Morris
Second Edition Copyright © 1982 Henry M. Morris

First Edition: *First Printing 1975*
                  Second Printing 1976
                  Third Printing 1977
                  Fourth Printing 1980
Second Edition: *First Printing 1982*

Published by
Creation-Life Publishers
P. O. Box 15666
San Diego, California 92115

First Edition ISBN 0-89051-015-6
Second Edition ISBN 0-89051-087-3
First Edition Library of Congress Catalog Card No. 75-15254
Second Edition Library of Congress Catalog Card No. 82-83647

Cataloging in Publication Data

Morris, Henry Madison, 1918 -
    The troubled waters of evolution. — 2d ed. —
San Diego : Creation-Life Publishers, 1982
    1. Evolution. 2. Creation. 3. Bible and evolution.
I. Title.
                         575   82-83647

Cover by Jay Wegter

Printed in the United States of America

CONTENTS

Table of Contents 1
Foreword 3
Introduction 5
Acknowledgments 7

CHAPTER I.   UP WITH CREATION!
The Premature Funeral of Creationism 9
Creation Revived by Scientists 13
Unresolved Problems of Evolution 16
Gaps in the Fossils 20
Is the Earth Really That Old? 20
The Faith of the Evolutionist 22

CHAPTER II.   TROUBLED WATERS EVERYWHERE
The Present Status of Evolutionary Thought 25
Anthropology, Geology, and Biology 28
The Physical Sciences 29
Psychology, Sociology, and Modern Education 30
History, Philosophy, and the Humanities 32
Ethics 34
Religion 37
Christianity 39
Totalitarian Ideologies 40
Racism 44
Control of Future Evolution 46

CHAPTER III.   A LONG, LONG TRAIL A'WINDING
The Darwinian Century 51
Why Darwin? 54
Before Darwin 62
The Pre-Christian World 65
The Beginning of Evolution 72

CHAPTER IV.   IN SCIENTIFIC CIRCLES
How to Define Evolution 79
Contrast with the Creation Model 82
Similarities, Differences, and Taxonomic Classification . 83
The Unprogressive Nature of Biologic Change 85
The Altogether Missing Links 88
The Order of the Fossils 92
The Law of Disorder 97
The Biblical Model of Creation 101

CHAPTER V.   **CAN WATER RUN UPHILL?**
Definitions 111
The First Law of Thermodynamics 114
The Second Law in Classical Thermodynamics 116
Entropy and Disorder 118
Information Theory and the Second Law 119
Are There Exceptions to the Second Law? 121
Criteria for a Growth Process 123
Trying to Save Evolution from Entropy 129
Origin of Matter and the Universe 132
Origin of Life 135
Origin of the Kinds 136
The Ubiquitous Entropy Principle 139

CHAPTER VI.   **BOIL AND BUBBLE**
Evolution and the Population Problem 145
Evolution, Energy, and Ecology 154
Evolution and Modern Racism 161
Evolution and the Sexual Revolution 166
Evolution and Life in Outer Space 168
Evolution and the Public Schools 171
Evolution Undermining Christian Education 178
Evolution Versus the Bible 184
Let God Be True 190

Index of Subjects 195
Index of Names 207
Index of Scriptures 210
For Further Reading 213

FOREWORD

Professor John N. Moore of Michigan State University tells about an occasion on which he had given his class an academically clear demonstration that the evolutionary viewpoint, with all its generalizations and speculation, is not scientific but, rather, is a religious position. One of the students expressed his objection to this conclusion by saying, "I believe in evolution!" Without realizing it this student provided yet another demonstration that evolution is a *belief*, or a faith. There are innumerable people today who are completely unaware of the fact that their belief in evolution is really a faith position, not a scientific position. This is as true for the scientist as for the nonscientist.

In my opinion, no one has done more in the last century to refute the false claims of that faith, that evolutionary dogma, than Dr. Henry M. Morris. His works are monumental: *The Genesis Flood* (with his collaborator, Dr. John C. Whitcomb, Jr.), *The Twilight of Evolution, The Remarkable Birth of Planet Earth, Scientific Creationism,* ten other books, and now this new book, *The Troubled Waters of Evolution.*

This book covers the historical background and modern influence of evolutionary thought in greater depth than ever before. The deadly influences of evolutionary philosophy are shown to be very real. As is true in all of Dr. Morris' books, there is extensive documentation. Evolutionary "science" is exposed for what it really is. The creationist alternative is presented clearly and shown to be fundamental and scientifically consistent with the best of science. The treatment is comprehensive. It is easy reading, informative, and the logic is delightful.

I never cease to be amazed at the skill with which Dr. Morris employs the writings of the top evolutionists themselves to develop an air-tight case against evolution. This book is filled with new source material of that type. The impact of the quotes and combinations of quotes from evolutionists in this book will surely cause more "troubled waters" for evolution when this book becomes widely read. One wonders how an evolutionist could know some of the things in these quotes and still be so foolish as to cling to the evolutionary position.

When students, who have been instructed in nothing but evolutionary doctrine in course after course, encounter such material as in this book, they soon see the sham in evolution and realize that special creation provides a better alternative. As teachers of science who hold to the Creation viewpoint, we continually hear these students raise the question,"Why haven't we been told this before?" It is the burning desire in Dr. Morris' life that all may have the opportunity to hear this and be brought to a knowledge of the historicity of the Scriptures, and of course to faith in the true God of Creation. Surely this book will help lead many to the light of truth.

I wish to express my personal gratitude to Dr. Morris for the example he has set in his own life and in his writings. It was his writings that first gave me inspiration and help to enter this "battle." It has been my great privilege to have known him as a Christian friend and co-worker in the Creation Research Society for many years. God has blessed America through this dedicated Christian scholar and works such as this book will be a blessing around the globe.

Thomas G. Barnes
Professor of Physics
University of Texas
El Paso, Texas
February 1975

INTRODUCTION

Until the last few years evolutionists have had everything going their way. Only a few ministers had courage enough to voice a dissent. Now for the first time, evolutionists are in trouble: they are confronted with men of equal competence in the fields of science, education, and experience who challenge them right on their home base, the college campus.

Dr. Henry Morris, director of the Institute for Creation Research, has been the most influential voice for creation, and has used science to confirm the Biblical explanation of man's origin. For six years he was president of the Creation Research Society, a group of over 700 creationist scientists holding Ph.D. or M.S. degrees. Dr. Morris' prolific writings have awakened many near-brainwashed educators and scientists to the fact that evolution cannot be supported by the facts of science. His books, *The Twilight of Evolution, The Genesis Flood, Scientific Creationism,* and many others have influenced thousands. What's more, he has the unique gift of being able to treat scientific material in a profitable and down-to-earth manner that everyone can understand and enjoy.

In this new book, Dr. Morris shows that the theory of evolution is the philosophical foundation for all secular thought today, from education to biology and from psychology through the social sciences. It is the platform from which socialism, communism, humanism, determinism, and one-worldism have been launched. All influential humanists are evolutionists, from Darwin, Huxley, Freud, and Pavlov to Rogers and Skinner. No one theory of man has ever influenced so many people. Accepting man as animal, its advocates endorse animalistic behavior such as free love, situation ethics, drugs, divorce, abortion, and a host of other ideas that contribute to man's present futility and despair.

The sad thing is that the theory of evolution is simply not true! In spite of the fact that it has wrought havoc in the home, devastated morals, destroyed man's hope for a better world, and contributed to the political enslavement of a billion or more people—it is only a delusion. Furthermore, it is the best financed and most overpropagandized delusion to confront man since Satan tricked Eve in the Garden of Eden.

The Troubled Waters of Evolution, reveals why and to what extent the evolutionists are in trouble. It may well prove to be the most amazing book you have read in the last decade!

Tim LaHaye
Author,
President, Family Life Seminars
June, 1975

ACKNOWLEDGMENTS

The writer would like especially to thank Dr. Tom Barnes and Dr. Tim LaHaye for reading the manuscript to *The Troubled Waters of Evolution,* and for writing the Foreword and the Introduction, respectively. In addition to being valued personal friends and colleagues, these two Christian gentlemen are recognized national leaders in two of the most vital movements in the nation today, both of which are directly relevant to the theme of this book.

Dr. Barnes is one of the nation's leading scientists in his own field of atmospheric physics, especially terrestrial electricity and magnetism. As President of the Creation Research Society (1973-1978), he is giving outstanding leadership to the scientific renaissance of creationism which is taking place in the nation today.

Dr. LaHaye, with his best-selling books on personal temperament and family living, and with his famous city-wide Family Life Seminars, has been spearheading a vital movement to restore the Bible-believing Christian family back to its rightful place in the lives of modern men, women, and young people.

Some of the material in this book was first developed for classroom use in the writer's course in "Scientific Creationism" at Christian Heritage College, and some was first prepared for publication in *Acts and Facts,* the monthly newsletter of the Institute for Creation Research. The remainder is presented herein for the first time. A number of students at the college and staff members at the Institute have helped in various ways.

As with most of his books, the manuscript was typed by Mrs. I. H. Morris, the writer's mother. At Creation-Life Publishers, General Manager George Hillestad and Editorial Manager Marilyn Hughes have been especially helpful in preparing the manuscript and selecting illustrations for publication.

For the past quarter century, a small group of scientists have been battling against the flood of evolutionary teaching released following the famous Tennessee "monkey trial" in 1925. In spite of the strength of the evolutionary establishment, these new scientific creationists are now making great numbers of converts to creationism among both scientists and laymen.

The Blowhole, Oahu, Hawaii

Photo credit: Col. M. G. McBee

CHAPTER I

UP WITH
CREATION!

The Premature Funeral of Creationism

The bells had tolled for any scientific belief in special creation. The Scopes trial (1925) had ended in a nominal victory for the fundamentalists, with the teacher Scopes convicted of teaching evolution in the high school, contrary to Tennessee law. In the press, however, Clarence Darrow and his evolutionist colleagues had resoundingly defeated William Jennings Bryan and the creationists. Evolution henceforth was almost universally accepted as an established fact of modern science, and special creation relegated to the limbo of curious beliefs of a former age.

A few states retained anti-evolution laws on their books for almost another generation, but these were no longer enforced. Most fundamentalists henceforth concentrated on "soul-winning" and "victorious living," keeping religion on a highly personalized and introspective basis and leaving the unfriendly realms of science and history to the intellectuals.

Some of the fundamentalists, of course, still realized it was necessary to fit the accepted geological history of the world and its inhabitants somewhere into their system. Bryan had relied heavily upon the "day-age" theory in his arguments at the Scopes trial, by which the days of Genesis were more or less equated with the ages of geology. Darrow and his witnesses had easily made this strained type of Biblical exegisis look ridiculous. In fact, Thomas Huxley had done the same thing half a century earlier, needling his clerical opponents with sarcastic remarks about a "Book which was so marvelously flexible that it could be made to mean whatever its users might wish it to mean!"

Most of the creationists retreated to the "gap theory," according to which all the geological ages and the fossil record associated with them were placed in a supposed time gap between the first two verses of Genesis. The six days of creation were, by this device, interpreted as six days of re-creation following a worldwide cataclysm which had terminated the pre-historic geological ages and left the pre-Adamic earth "waste and void" (Genesis 1:2).

Since the geological age-system and its corresponding successions of life in the fossil record constitute the chief evidence for evolution, the gap-theory, of course, provided no answer to evolution, as many fundamentalists thought — it merely pigeonholed it in the so-called "gap." Since there was no geological evidence for any such pre-Adamic cataclysm, the scientists henceforth simply ignored the gap-theory creationists and proceeded to teach and apply the evolutionary philosophy, unrestrained, in every field. In the past half-century evolutionary thought has permeated literally every discipline, from biology to astronomy and from psychology to political science. Evolution has, since that time, been taught in the public schools in one way or another from kindergarten to graduate school, and any lingering doubts by teachers or pupils are quickly laughed off as unscientific.

Sometimes, evolution is described as God's "method of creation," in an attempt to make it more palatable to die-hard creationists, but this device has never been satisfactory, either to evolutionists or creationists. Scientific evolutionists contend that, since mutation and natural selection are sufficient to explain evolution, no supernatural designer is needed to initiate or to operate the evolutionary process. At the same time, creationists argue that a divine Creator would never invent such a cruel and inefficient process as evolution to "create" man, if He is really the God of love and power the Bible represents Him to be.

Thus, since a system of theistic evolution was not wanted or needed by the evolutionists and since the fundamentalists were unable to develop a satisfactory theory to harmonize Genesis with geology, it seemed that creationism at last was dead. Many people even proclaimed that God Himself was dead!

Evolutionists were not satisfied however. Although creationism was no longer taught in the schools or included in the textbooks, they complained that there was not *enough* evolution being taught! For a while after the Scopes trial, textbook publishers were sensitive to the strong tide of creationist belief which had been exhibited for several years prior to the trial and so soft-pedalled the treatment of evolution in a number of their books.

"The impact of the Scopes trial on high school biology text-books was enormous. It is easy to identify a text published in the decade following 1925. Merely look up the word 'evolution' in the index or the glossary; you almost certainly will not find it." [1]

Evolution was always assumed or implied, however, even if somewhat veiled. In the meantime, the fundamentalists had more or less retreated from the conflict, so the evolutionary scientists, who had long since gained almost full control of higher education, soon sought to consolidate and strengthen their dominance over the public schools.

"The corresponding group of biologists, the American Institute of Biological Sciences, produced the texts known as the Biological Sciences Curriculum Study (BSCS) texts; these completely transformed the profile of high school biology texts." [2]

Not only the evolutionary biologists were behind the new BSCS books, however.

---

[1] Judith V. Grabiner and Peter D. Miller: "Effects of the Scopes Trial," *Science,* Vol. 185, September 6, 1974, p. 833.

[2] *Ibid.,* p. 836.

"The prestige, power, and financial support of the federal government were behind the scientists and the new textbooks."[1]

An $8,000,000 grant was obtained from the National Science Foundation to get the books written, and a BSCS Center was set up at the University of Colorado to implement their publication and adoption. The BSCS books were, of course, all thoroughly saturated with evolutionary philosophy, though to different degrees in different "versions."

In the meantime, the beautifully illustrated Time-Life books on evolution were going into millions of homes and school libraries; television specials were emphasizing evolution, and the courts were banning prayer and the Bible from the public schools. Obvious implications were being widely drawn from the now-accepted "fact" of man's bestial ancestry, and young people (along with not a few of their elders) were being conditioned by the movies and the popular literature for a rapid descent into radicalism and amoralism.

Evolutionists by and large became smug, secure in their confidence that evolution was now firmly established as the official "religion" of the state and of society in general.

The high point of their triumph probably was the great Darwinian Centennial Convention at the University of Chicago in 1959. There the world's leaders of evolutionary thought gathered together to pay homage to Charles Darwin, on the one-hundredth anniversary of the publication of *Origin of Species,* spending an entire week in giving papers and discussions on the myriad applications and implications of evolution in the modern world. Sir Julian Huxley was the keynote speaker, and he summarized his views for the press in words of choice arrogance:

"In the evolutionary system of thought there is no longer need or room for the supernatural. The earth was not created; it evolved. So did all the animals and plants that inhabit it, including our human selves, mind and soul, as well as brain and body. So did religion.

"Evolutionary man can no longer take refuge from his

---

[1] *Ibid.*

loneliness by creeping for shelter into the arms of a divinized father figure whom he himself has created."[1]

Many like assertions were made by Huxley and the other speakers that week. Not once, so far as the record of the proceedings goes, did any speaker or discusser or questioner from the audience support creationism or even theistic evolution, nor did anyone raise any objections to any of these atheistic pronouncements by Huxley and his colleagues. Creationism and the Creator seemed to be dead and buried, once and for all.

Creation Revived by Scientists

But if creationism once was dead, it has recently risen from the dead! Today there are hundreds of outspoken scientists advocating a return to creation and abandonment of evolution, and their numbers are increasing. The evolutionary "establishment" is becoming alarmed, as multitudes of disillusioned youth are recoiling from the precipice of animalistic amoralism and survival-of-the-fittest philosophy to which two generations of evolutionary indoctrination had led them.

Actually, creationism had not really died, though it had to go mostly underground. During the 1930's and 1940's, it would have been difficult to find any scientist at the state colleges and universities who would advocate creationism. A few writers on the faculties of fundamentalist schools continued to oppose evolution. In England, the Evolution Protest Movement included in its leadership a number of recognized scientists, as did the short-lived Society for the Study of Creation and the Deluge in this country.

There were a few writers during this period who kept opposition to evolution alive among fundamentalists. These included Harry Rimmer, Arthur I. Brown, George McCready Price, and others.

Two developments during the 1940's were significant, the formation of the American Scientific Affiliation and the Moody Institute of Science. Although neither of these organizations was overtly anti-evolutionist, they did stress the innumerable

---

[1] Associated Press dispatch, November 27, 1959.

evidences of design in nature and the necessity of interpreting science in a theistic context. Both were conservative and Bible-centered and yet were genuinely scientific organizations. The A.S.A. is a society of evangelical scientists, several hundred strong. Although most of them were and are theistic evolutionists, and few were among the scientific leaders in the state universities, the Affiliation did provide a sort of haven and rallying ground for conservative Christians in the scientific world, as well as a journal and meetings devoted to a continuing analysis of the relation between science and Scripture.

The Moody Institute of Science (Whittier, California) is widely known for its outstanding series of filmed "Sermons from Science." These skillfully-produced and beautiful films have been viewed by millions in the armed services, at world fairs, in the schools, and at many other places. None of these films is explicitly anti-evolutionary, but they all point out the overwhelming necessity for an omnipotent and omniscient Creator behind the universe and have convinced hosts of young people and others that the Bible does after all still possess a high degree of scientific validity.

These developments, and a number of influential books authored by men in these movements, finally culminated in the formation in 1963 of an aggressively creationist scientific organization. The Creation Research Society, organized in 1963 by ten scientists, now has a scientific membership of over 700 voting scientist members and a lay associate membership of approximately 2000. To be a voting member, one must have either a Master's degree or a Ph.D. in one of the natural sciences and must also subscribe to a Statement of Faith which includes commitment to belief in special creation and a worldwide Noahic flood, as well as opposition to any form of evolution—including theistic evolution!

The Society publishes a quarterly journal[1] of scholarly, well-documented articles on scientific creationism and catastrophism.

---

[1] For information regarding this journal or membership in the Society, write its Managing Editor, Dr. John N. Moore, Professor of Natural Science at Michigan State University in East Lansing.

Many of its scientist members were evolutionists at one time, but have become convinced that creation is a more reasonable scientific explanation of origins than evolution. The Society in late 1970 published an attractive and comprehensive high school biology textbook, the first such book in which the factual data of biological science are structured around creation rather than evolution.

Since the formation of the Creation Research Society in 1963, there has been a great proliferation of other creationist membership associations. Among the national organizations in this country are the Bible-Science Association (headquartered in Minneapolis), the Creation Social Science and Humanities Society (Wichita), and Students for Origins Research (Santa Barbara). Statewide creationist organizations have been formed in Missouri, Ohio, and many other states, not to mention large numbers of local citizens' groups.

In recent years, national creationist organizations have been formed in many other countries as well. Probably the first was the Creation Science Association of Canada. Two of the more active organizations are in England and Australia, but there are also strong groups in Korea, Sweden, Holland, Brazil, India, Nigeria, Japan, and other countries. All of these have placed special emphasis—sometimes exclusive emphasis—on the *scientific* superiority of creationism over evolutionism.

The organization which has had the greatest impact for creationism, however—at least according to most concerned evolutionists—has been the Institute for Creation Research. ICR is not a membership organization like the others, but an actual educational/research/publishing organization, with a staff of 12 Ph.D. scientists, plus supporting staff. Organized only in 1970, the Institute has developed over 60 books, given lectures and messages in many hundreds of churches and on other hundreds of college and university campuses, conducted a weekly worldwide radio program, and participated in a wide variety of evangelistic and teaching ministries. Of greatest influence have been the creation-evolution debates between ICR scientists and leading evolutionary scientists. Almost 150 of these have been held to date (1982), usually before audiences numbering in the thousands. Since 1981 ICR has offered graduate degree programs in the sciences. Its worldwide impact has been far greater than can be attributed to its very limited size and resources, so its staff and supporters are convinced that God Himself is leading and using its work.

Thus creationism is well on the way back (even though the evolutionary opposition has become very bitter), this time not only as the teaching of the Bible, but also as a superior scientific explanation of the world in which we live.

Unresolved Problems of Evolution

Modern creationists recognize and accept all the observed biological changes which evolutionists offer as proof of evolution. New varieties of plants and animals can be developed rather quickly by selection techniques, but creationists point out that no new basic *kind* has ever been developed by such processes. Mutations are fairly common, but not transmutations. A moth species may change from predominantly light-colored to predominantly dark-colored, as a result of natural selection operating on Mendelian variants in a changed environment, but the moth does not become a dragon-fly, or even a different species of moth! Fruit-flies may develop numerous new mutants as a result of irradiation, but after a thousand successive generations of such treatment, they are still fruit-flies!

Evolutionists, however, object to this argument, insisting that the few thousand years of recorded history are insufficient time to permit new kinds to evolve. For this, millions of years are required, and so we should not expect to *see* new kinds evolving.

Creationists in turn insist that this belief is not scientific evidence but only a statement of faith. The evolutionists seems to be saying: "Of course, we cannot really *prove* evolution, since this requires ages of time, and so, therefore, you should accept it as a proved fact of science!" Creationists regard this as an odd type of logic, which would be entirely unacceptable in any other field of science.

As a matter of fact, it is not true that time alone will produce evolution. The common opinion that anything can happen, no matter how improbable, if enough time is available, is obviously incorrect. For example, a pile of bricks and lumber would never "evolve" into a building by the random interplay of environmental forces acting upon it, no matter how many billions of years it might lie there on the construction site. On the contrary, it would beyond question "*de*-volve," and go back to dust.

Here it is necessary to introduce the concept of entropy, which creationists insist makes evolution on any significant scale quite impossible. Entropy (from two Greek words meaning "turning inward") is a measure of disorder, or randomness. The Law of Increasing Entropy, known also as the Second Law of Thermodynamics, states that *every* system (physical, biological, or anything else) tends toward a state of increasing entropy.[1] This is why machines run down, why houses and roads wear out, and why organisms get old and die. As time goes on, disorder increases. The great British scientist, Sir Arthur Eddington, called this law Time's Arrow. That is, as time flows onward, the "arrow" of available energy for future processes always points downward. The whole universe, in fact, is "running down," heading toward an eventual "heat death," in which all forms of energy will have been degraded into uniform, low-temperature heat dispersed through space, no longer capable of accomplishing the work of maintaining the innumerable processes of the cosmos.

Creationists point out that this Second Law, accepted by all scientists and confirmed by thousands of experiments on all types and sizes of systems, is exactly the converse of the principle of evolution, according to which everything is moving upward from primeval chaos to future perfection. Therefore, they ask, how can evolution and entropy both be universal laws? But there is no doubt that the Second Law has always been confirmed quantitatively, wherever it can be tested.

It is true, of course, that there can be *apparent* exceptions to the Second Law, in so-called "open systems." Thus, a seed may

---

[1]See Chapter V for a thorough discussion of this very important fallacy in evolutionary theory.

grow up into a tree or a pile of bricks and lumber into a building. An outside source of energy is necessary for this to occur — sunlight in the case of the seed, construction machinery and workmen in the case of the building. But even then, this apparent growth is only local and temporary, not universal and eternal. *Most* seeds die, and even the few that become trees also *eventually* die. The building, too, will eventually go back to the dust.

The evolutionist agrees, but he also says that, since the earth is an open system, there is enough incoming solar energy to maintain the evolutionary process over the "few" billion years of geologic time before the earth finally dies. It may be, some also say, that even though all processes now tend toward increasing entropy, perhaps in the past they were different.

But again, to the creationist, this is a strange type of reasoning for scientists. Science is supposed to proceed from confirmed facts of experimental observation. How can processes which now *always* involve *increasing* entropy justify belief in past evolutionary processes of *decreasing* entropy?

Furthermore, even local and temporary increases in order require more than merely an open system with energy available in the environment. *Always,* in addition, there must be a program (or code, or pattern, or template) built into the embryonic system to direct the conversion of the environmental energy into meaningful growth of order in the system. In the case of the seed, this requirement is met in part by the marvelous genetic code, the intricately-ordered structure of the DNA molecules of the genes of the germ cell. This amazing biochemical system could never have created itself and yet it is absolutely necessary for the system to grow and it always directs the inflowing energy into the production of a plant (or animal) of the same basic kind as the parents — even when mutational "mistakes" in the transmission of this information result in somewhat deformed or otherwise damaged offspring.

The question is: "What, or where, is the infinitely complex program which would be required to convert the primeval disorder into the infinitely-complex ordered structure of the universe and its living creatures?" The evolutionist at this point pleads ignorance, saying such questions are not proper questions for science to ask. The creationist, of course, says the logical answer is an infinitely-capable Programmer! But if such an Infinite Intelligence exists, then He would be too intelligent to create man by such an infinitely cruel, wasteful, inefficient process as evolution!

In addition to a program, any system experiencing a growth of order must also involve an energy conversion of some kind for transforming the environmental energy into the growth phenomenon. For the seed, this is the amazingly-complex, little-understood process known as *photosynthesis*. This process could never create itself, or just randomly happen, but there it is! Order never spontaneously arises out of disorder.

It is not sufficient to say, as the evolutionist does, that the sun's energy is great enough to maintain the process of evolution. The essential, and unanswered, question is: "*How* does the sun's energy maintain the process of evolution? What is the specific mechanism of 'evolutionary photosynthesis' that converts solar energy into the transformation of particles into atoms, then into molecules and stars and galaxies, complex molecules into replicating molecules, simple cells into metazoan life, marine invertebrates into reptiles and birds and men, unthinking chemicals into conscious intelligence and abstract reasoning?" Again the creationist is convinced that special creation is a more scientific theory than evolution. He notes that the pile of bricks and lumber will never evolve by itself into a building, even though it also is an "open system" and the sun's energy bathing the construction site contains far more than enough energy to carry out the process of building the building. In fact, the building is far less complex than even the simplest form of living cell, not to mention the total organic world itself.

The evolutionist may somehow maintain his faith that the phenomena of mutation (seemingly a disordering mechanism, in full accord with the Second Law) and natural selection (seemingly a conservational process which tends to weed out the misfits produced by mutation, thus maintaining the *status quo* in nature) supply the mechanism for evolving the whole world, but he ought to recognize that this is pure faith (or, better, wishful thinking), not science!

Gaps in the Fossils

But, at this point the evolutionist reminds the creationist of the fossil record and the long geological ages. The fossilized remains of plants and animals preserved in the sedimentary rocks of the earth's crust are cited to prove the historical *fact* of evolution, even though the precise mechanism may be obscure. Extinct trilobites and dinosaurs and cavemen presumably witness to the tortuous evolutionary struggle over hundreds of millions of years of geologic time. Furthermore, radioactive dating by uranium, potassium, and other minerals confirms the evolutionist's belief that the earth must be billions of years old.

But this evidence does not persuade the creationist either. He reminds the evolutionist that *extinction* does not prove *evolution*! The fossil record speaks eloquently of death, often catastrophic death. The fact that dinosaurs have become extinct tells us nothing about how they first came into existence.

Furthermore, essentially the same gaps that exist between basic kinds in the present world exist also in the fossil world (e.g., the gap between cats and dogs, pines and oaks, sharks and whales, etc.) Multitudes of fossils exist, representing the same kinds of animals as in the present world (as well as extinct kinds, such as dinosaurs and trilobites), but none has been found in between these basic kinds, or leading up to them! This strikes the creationist as a highly improbable circumstance if indeed all basic kinds have evolved by slow stages from a common ancestral form, as the evolutionist insists. There ought rather to be a continuous intergrading series, instead of discrete kinds, both in the present world and in the fossils, if evolution is really true.

Is the Earth Really That Old?

Neither is the modern scientific creationist much impressed by the supposed great age of the earth's geologic formations. He reminds us that the only real *history* that has been recorded from earlier times extends back just a few thousand years. Any events that took place before that must be deduced indirectly by extrapolating some present-day process (e.g., radioactive decay of uranium into lead, soil erosion, canyon-cutting, delta deposition, etc.) back into the distant past, to arrive at an approximate date when those events took place.

But to do this, several assumptions have to be made: (1) the condition of the components of the process when it first began to operate has to be known (but this is impossible, since no observers were present then); (2) the system must always have been a "closed system," so that no external events could affect the process (but this is impossible, since there is in nature no such thing as a truly closed system); (3) the assumption of *uniformitarianism* must be applied, by which the process is assumed always to operate at the same uniform rate, throughout all ages (but this is impossible, since all processes are statistical in nature, and their rates can and do vary and change whenever any of the many factors controlling them vary and change).

Furthermore, creationists object to the screening process by which evolutionists discard all natural processes which do *not* give great ages for the events of earth history. There are many such processes (e.g., influx of helium into the atmosphere, influx of uranium into the ocean, influx of meteoritic dust to the earth, and many others) which indicate the earth is quite young! Even many uranium and potassium ages turn out to be practically zero, but these are invariably discarded or explained away on the basis of supposed contamination or alteration.

Still further, the creationist suspects that the fossil record and the sedimentary rocks, instead of speaking of a long succession of geologic ages, may tell rather of just *one* former age, destroyed in a single great worldwide aqueous cataclysm. Several objective facts suggest this: (1) the fact that the geologic "ages," as identified primarily by their fossils, are found in quite an indiscriminate variety of arrangements at different localities around the world, with any succession possible — any "age" missing, or even laid on top of any supposedly more "recent" age; (2) the fact that rocks of any physical character — limestone, shale, sandstone, metamorphic, igneous, loose and unconsolidated, hard and indurated, etc. — may be found indiscriminately in any so-called "age"; (3) the fact that any given local formation seems almost always to indicate geologic processes (e.g., volcanism, glaciation, tectonism, sedimentation, etc.) operating at far higher, more catastrophic intensities than is the case in the modern world; (4) the fact that fossils are often, in all "ages," found buried together in great numbers, under conditions absolutely requiring cataclysmic extinction and rapid burial; (5) the fact that the traditions of all nations and tribes around the world describe just such a primeval watery cataclysm as the geologic formations seem to require.

The Faith of the Evolutionist

The evolutionist may finally admit that his theory does still have many serious unsolved problems. Nevertheless, he feels it is the only proper belief, since belief in special creation in effect gives up on the problems, relying on a force outside present scientific phenomena to explain the origin of these phenomena.

The creationist acknowledges this. He finally must accept creation and a Creator by faith, since the process of special creation is not accessible to scientific observation.

But neither is the historical process of evolution, he reminds the evolutionist. Evolution also must be accepted on faith, and that faith is more arbitrary than that of the creationist. Evolutionary faith must be maintained in spite of the Second Law of Thermodynamics, the clearcut distinction between kinds, the very limited nature of observed biological changes, the deteriorative nature of mutations, the many contradictions in the fossil record, the catastrophic appearance of most geologic formations, and many other problems. These phenomena are all perfectly consistent with creationism, of course.

In any case, creation at the very least has as much plausibility as a scientific model of origins as does evolution. Therefore, creationists are saying with increasing clarity these days, creation should be taught in the public schools on at least an equal basis with evolution.

The idea of evolution is not merely a certain biological theory having to do with men and monkeys, but is a complete world view, a philosophy of life, the established religion of the state. Its pervasive influence has penetrated every field of study and has provided the pseudo-scientific basis of communism, fascism, racism, animalism, and all the other deadly philosophies that trouble the world today.

Mediterranean Sea

Photo credit: Tim Ravenna

CHAPTER II

TROUBLED WATERS EVERYWHERE

The Present Status of Evolutionary Thought

For several decades after the Scopes Trial, most people, including even most Bible-believing Christians, tended to regard the evolution-creation issue as one of very little importance. The question of uniformitarianism versus catastrophism, including whether the Flood of Noah was worldwide or only local, was considered of even lesser importance.

These events of the far-off past, many felt, have no relevance to the pressing problems of the modern world, and thus were matters of little concern to them. Both questions could well be left to the professional geologists, in the case of uniformitarianism, and to the biologists, in the case of evolutionism. Even in the matter of Christian faith, the gospel of Christ does not depend upon any particular cosmological model; both evolutionists and creationists can believe in God and can be saved through a personal faith in Jesus Christ, so the argument went. A person's relation to God and to his fellow-men did not depend upon whether or not he believed the Flood was worldwide. If necessary, they felt, it would somehow be possible to reinterpret Genesis to correspond to the prescriptions of the evolutionary scientists without any particular damage to the basic Biblical philosophy.

It is this attitude of indifference that is, if anything, more harmful to true faith, however, than even open hostility by the camp of atheism. It *does* make a tremendous difference what men believe about their origin, and the sad history of the Christian church of the past 150 years ought to be sufficient proof of this fact. The evolutionary-uniformitarian cosmology is far more than a mere biological or geological hypothesis. It is a complete world-view, a philosophy of life and meaning. One cannot really believe in an evolutionary history of the world without also believing in an evolutionary future of the world. His philosophy of origins will inevitably determine sooner or later what he believes concerning his destiny, and even what he believes about the meaning and purpose of his life and actions right now in the present world.

The doctrine of origins, indeed, is the foundation of every other doctrine. That, of course, is why God placed His revelation of origins in the first chapter of the Bible. Everything else in the Bible and in history is built upon this foundation. Once the foundations have been undermined, of course, it is only a matter of time before the entire superstructure must collapse. The Harvard biologist, Dr. Ernst Mayr, one of the world's leading evolutionists, in an important recent article, confirms this:

"I am taking a new look at the Darwinian revolution of 1859, perhaps the most fundamental of all intellectual revolutions in the history of mankind. It not only eliminated man's anthropocentrism, but affected every metaphysical and ethical concept, if consistently applied."[1]

With respect to God's role in the universe, the evolution model essentially eliminates it altogether (as practically all of the leaders of evolutionary thought believe) or at least relegates it to some

---

[1] Ernst Mayr: "The Nature of the Darwinian Revolution," *Science,* Vol. 176, June 2, 1972, p. 981.

external and innocuous supervision which can never be demonstrated or even studied in any scientific way. Mayr continues:

"Every anti-evolutionist prior to 1859 allowed for the intermittent, if not constant, interference by the Creator. The natural causes postulated by the evolutionists completely separated God from his creation, for all practical purposes. The new explanatory model replaced planned teleology by the haphazard process of natural selection. This required a new concept of God and a new basis for religion." [1]

Because it is important for people to face the fact that the question of origins really *is* a most vital issue that can be ignored only at great peril, this chapter will be devoted to demonstrating that evolutionary theory has permeated every field of thought and study, and is having a profoundly deleterious effect everywhere. More and more people, in fact, are today becoming aware and concerned about this matter. This is true especially among Christian young people, who have sensed the failures and dangers of the evolutionary philosophy which permeates the "Establishment" today in almost every field.

At present, evolutionary thought is dominant not only in biology but in all other disciplines as well. The creationist cosmology has been held by only a small minority and too few of these have had any substantial scientific comprehension of its implications.

A leading evolutionist emphasizes the universal scope of the evolutionary process as follows:

"Evolution comprises all the stages of the development of the universe: the cosmic, biological, and human or cultural developments. Attempts to restrict the concept of evolution to biology are gratuitous. Life is a product of the evolution of inorganic nature, and man is a product of the evolution of life." [2]

A widely-used university textbook, though less dogmatic than others in this field, begins with the words:

"Organic evolution is the greatest general principle in biology.

---

[1] *Ibid*, p. 988.
[2] Theodosius Dobzhansky: "Changing Man," *Science,* Vol. 155, January 27, 1967, p. 409.

Its implications extend far beyond the confines of that science, ramifying into all phases of human life and activity. Accordingly, understanding of evolution should be part of the intellectual equipment of all educated persons."[1]

Anthropology, Geology, and Biology

The accepted theory of man's development leaves no room for the Biblical Adam, but rather visualizes a gradual divergence of man and the apes from a common ancestor in the Paleocene Epoch about 30 to 60 million years ago, with man finally evolving into essentially his present form perhaps four million years ago. Human evolution is insisted upon as factual, not even open to question.

"That man has evolved from less distinguished ancestors is indisputable. What we are concerned with is not to show where man came from. That we no longer doubt. But to show how he came; to show the processes by which ape-like animals became men."[2]

Historical geology attempts to decipher the history of the earth; its entire theoretical structure is based upon evolution, as interpreted from the fossils found in the sedimentary rocks of the crust. The geologic time scale, which is the backbone of all geologic interpretation, is based squarely on the theory of evolution. The University of California paleontologist, W.B.M. Berry, has written an entire book with this primary theme. He says:

"This book tells of the search that led to the development of a method for dividing pre-historic time based on the evolutionary development of organisms whose fossil record has been left in the rocks of the earth's crust."[3]

It hardly is necessary to point out the profound influence of evolution on the study of life processes. It is here, in biology, that organic evolution is best known. All standard biology textbooks

---

[1] Paul A. Moody: *Introduction to Evolution* (2nd Ed., New York, Harper and Row, 1962), p. 1x.

[2] C. D. Darlington: *The Evolution of Man and Society* (New York 1970), pp 32, 21.

[3] W.B.N. Berry: *Growth of a Prehistoric Time Scale* (San Francisco, W. H. Freeman Co., 1968) p.v.

today, with one pioneering exception,[1] are structured entirely around the evolutionary model, and this is true at every grade level. Life is always presented as having evolved by natural processes from non-life, all the various higher forms of life from simpler forms, and finally man himself as the highest product of the evolutionary process to date. Whether the particular study is botany, zoology, genetics, ecology, embryology, or any of the other many branches of biological science, the underlying philosophy is always that of evolution.

The Physical Sciences

The physical sciences are not as much affected by evolutionary thinking as are the life sciences. Today, however, even physics is influenced a great deal by cosmogonic speculations.

"Modern astrophysics has brought a new aspect to physics: the historical perspective. Previously, physics was the science of things as they are; now, astrophysics deals with the development of stars and galaxies, with the formation of the elements, with the expanding universe . . . stars are formed from a hydrogen cloud, elements are formed by synthesis from hydrogen, and stars are developing through different states."[2]

Even in such fields as mathematics and technology, the authors of textbooks commonly feel it necessary to begin their treatments with their own purely imaginary speculations about how their particular field of technology first evolved. For example, a recent mathematics book begins thus:

"In the beginning, there were no numbers; or, if there were, primitive man was unaware of them. Whether the numbers were always 'there' (where?) or had to be invented, has been a much discussed question, and we shall leave it to the philosophers to continue that discussion without our aid. What

---

1 *Biology. A Search for Order in Complexity,* Ed. by John N. Moore and Harold S. Slusher (2nd Ed., Grand Rapids, Zondervan Publishing House, 1974). This book was prepared by 20 scientists of the Creation Research Society, most with doctorates in biology.

2 Victor K. Weisskopf: "Physics in the Twentieth Century," *Science,* Vol. 168, May 22, 1970, pp. 929-930.

we can say with some assurance is that the ability to count came relatively late to civilization."[1]

Psychology, Sociology and Modern Education

Turning to the social sciences, it is apparent that these have been more influenced by evolutionary thought than even the biological sciences. Psychology, the study of the mind, is almost entirely based on the assumption that man is only an animal, derived by evolutionary descent from an ape-like ancestor. This, of course, was the view of Sigmund Freud, who has exerted probably the greatest single influence on the structure of modern psychology. It is a common saying that, as Darwin banished God from life, Freud drove Him from the soul. Freud's proverbial emphasis upon the "unconscious" and on uninhibited sexual freedom are both based squarely upon man's supposed brute ancestry.

The same can be said in varying degrees about the contributions of James, Watson, Jung, and other founders and leaders of psychology, including the so-called humanistic psychologists of the present day. Dr. Rene Dubos, of the Rockefeller Institute, in a national Sigma Xi — Phi Beta Kappa lecture, noted this as follows:

"Many aspects of human behavior which appear incomprehensible, or even irrational, become meaningful when interpreted as survivals of attributes which were useful when they first appeared during evolutionary development and which have persisted because the physical evolution of man came to a relative halt about 150,000 years ago. Phenomena ranging all the way from the aberrations of mob psychology to the useless disturbances of metabolism and circulation which occur during verbal conflicts at the office or at a cocktail party are as much the direct consequences of the stimuli which were their immediate causes. The urge to control property and to dominate one's peers are also forms of territoriality and dominance among most if not all animal societies."[2]

[1] C. S. Ogilvey and John L. Anderson: *Excursions in Number Theory* (New York, Oxford University Press, 1966), p. 1.
[2] Rene Dubos: "Humanistic Biology," *American Scientist,* Vol. 43, March 1965, pp. 10-11.

Even the apparent intelligence and purposeful actions which are characteristic of man are believed to be merely products of the random, purposeless evolutionary process. Dr. Hudson Hoagland, then President of the American Academy of Arts and Sciences, said for example:

"But man himself and his behavior are an emergent product of purely fortuitous mutations and evolution by natural selection acting upon them. Non-purposive natural selection has produced purposive human behavior . . . ." [1]

Such writers rarely bother to concern themselves with the obvious contradiction of these concepts with the scientific law of cause and effect, since they simply regard evolution as having in some mysterious way transcended this law. In any case, with the exception of a very small minority today, the professions of psychology and psychiatry are characterized by a strong commitment to evolution and animosity to Biblical Christianity.

Turning to the field of sociology, one quickly discovers that the study of man's cultures and societies is universally cast in the same mold as the study of his presumed biological evolution.

"A second kind of evolution is psycho-social or cultural evolution. This is unique to man. Its history is very recent; it started roughly a million years ago with our hominid tool-making ancestors . . . In the last 300 years the ever-accelerating developments through science are a continuation of this psychosocial evolution, which, in terms of progress, is thousands of times faster than biological evolution resulting from genetic mutations." [2]

Similarly, one of the leading social scientists of the past quarter century, Dr. Harry Elmer Barnes, has said:

"Unquestionably the most potent influences contributing to the rise and development of truly historical sociology were Spencer's theory of cosmic evolution and the Darwinian doctrine of organic evolution and their reactions upon social science." [3]

[1] Hudson Hoagland: "Science and the New Humanism," *Science,* Vol. 143, January 10, 1964, p. 113.
[2] *Ibid.*
[3] Harry Elmer Barnes: *Historical Sociology,* (New York, The Philosophical Library, 1948), p. 13.

More recently, Alan Lomax, of the Bureau of Applied Social Research at Columbia University, concluded an important worldwide study of cultural characteristics as follows:

"For almost a century, the intellectual atmosphere of the world has been poisoned by a false Darwinism that judged human social development as the survival of the fittest — that is, of the most successfully aggressive individuals and societies. This view can now be corrected."[1]

It is a matter of increasing concern to Christian parents and pastors that evolution is now taught as fact in practically all public schools and at all grade levels. Science and social studies textbooks invariably support evolution. It is encouraging that organized opposition to this clearly unfair and unconstitutional state of affairs has developed in practically every state, with strong citizens' groups advocating a return to a more balanced approach to the study of origins. It is not as well-known, however, that not only in organized courses, but even in the very philosophy and methodology of the entire public education approach, evolutionary assumptions have been at the foundation. The curricular content, the "self-discovery" emphasis on social change, and many other aspects of modern educationism are founded on evolutionary assumptions. The determinative influence of John Dewey and his disciples in modern education of course speaks for itself.

History, Philosophy and the Humanities

Modern textbooks of world history invariably [2] begin with the standard evolutionary development of the world and man, supplemented by the evolutionary interpretation of developing nations, races, and classes. This kind of approach is of course made to order for the Marxist and Nazi distortions of history, as will be discussed shortly. Even apart from these, however, historical interpretations today are thoroughly permeated with

[1] Alan Lomax, with Norman Berkowitz: "The Evolutionary Taxonomy of Culture," *Science,* Vol. 177, July 21, 1972, p. 239.

[2] Again there is one exception, *Streams of Civilization,* by Albert Hyma and Mary Stanton: (San Diego, Creation-Life Publishers, 1976). This book is a high school world history textbook, and is the only such book, so far as we know, which is creationist and Biblical in its orientation.

evolutionary assumptions. Thus, H. J. Muller, in speculating about pre-historical cultural development, said:

"Another factor that must have facilitated cultural evolution in the past is a kind of non-genetic natural selection operating between different groups, and between portions of a group, so as to favor more the continuance and spread of those whose cultures were more conducive to their own survival and increase."[1]

It is, of course, well-known that the most influential of our modern historians, men such as Charles Beard and Arnold Toynbee, have uniformly been committed to doctrinaire evolutionary concepts of history, both ancient and modern.

There are, of course, a great many different philosophies, but all of them — except a genuine Biblical philosophy — are evolutionary philosophies. Most of those now prominent are essentially variant forms of naturalism. The philosophies of Karl Marx and Friedrich Nietzsche — the forerunners of Stalin and Hitler — have been particularly baleful in their effects. Both were dedicated evolutionists. Perhaps the most influential American philosopher was John Dewey, and his philosophy also was built on Darwinism and a sort of pantheistic humanism.

"Dewey was the first philosopher of education to make systematic use of Darwin's ideas."[2]

The humanities include literature and the fine arts, as well as philosophy and history. There is such a gulf between the sciences and the humanities that they have actually been widely referred to as the "two cultures." However, in terms of the basic world views which they hold, they are one. The humanities, no less than the natural sciences and the social sciences, regard man as the naturalistic product of his environment. The philosophy, morality, esthetics, and other aspects of modern humanism all are rooted in naturalism and evolutionism.

"Evolutionary concepts are applied also to social institutions and the arts. Indeed, most political parties, as well as schools

---

[1] H. J. Muller: "Human Values in Relation to Evolution," *Science,* Vol. 127, March 21, 1958, p. 628.

[2] Christian O. Weber: *Basic Philosophies of Education* (New York, 1960), p. 252.

of theology, sociology, history, or arts, teach these concepts and make them the basis of their doctrines. Thus, theoretical biology now pervades all of western culture indirectly through the concept of progressive historical change."[1]

One need only explore modern literature, listen to modern music, watch modern drama, or view modern art to become quickly convinced that they are all pervaded by a spirit of amoralism and atheism that can only be grounded in the belief that science has proved man is an animal and God is dead.

Ethics

Once man has rejected the Bible and other religious authority, there is no more divine constraint toward honesty or purity or charity or any of the other ethical values associated with divine revelation.

"An ethical system that bases its premises on absolute pronouncements will not usually be acceptable to those who view human nature by evolutionary criteria."[2]

Nevertheless, man cannot survive in chaos and anarchy, and therefore must have some kind of ethical standard. Since "science" has taken away his former guide, many scientists feel constrained to provide, in substitution, a new "scientific ethics," and the scientific journals frequently carry articles devoted to this theme.

And what is the basis of this scientific ethics? Why, evolution, of course!

"Moreover, scientific thought and its devotion to truth are themselves a product of the evolutionary process, and must therefore have proved themselves, hitherto at any rate, to have survival value. But the idea that all ethical considerations can be derived from the evolutionary process has not gone unchallenged."[3]

Which latter fact is, at least, somewhat reassuring. However, there does seem to be an increasing clamor from evolutionary

---

[1] Rene Dubos, *op cit,* p. 6.

[2] Arno G. Motulsky: "Brave New World?" *Science,* Vol. 185, August 23, 1974, p. 654.

[3] Walter R. Brain: "Science and Antiscience," *Science,* Vol. 148, April 9, 1965, p. 197.

scientists that they could and should direct future evolution and human societies in accord with their concept of a scientific evolutionary system of ethics, H. J. Muller said:

"It is high time for modern man, everywhere, again to revise his concepts of values, in accord with the utterly new view that science, and especially evolutionary science, has given him of the nature of the world and of his actual and potential relations to it."[1]

Seventy-five years ago, John Dewey, the architect of our modern system of public education, made a profoundly influential address on the subject of evolution and ethics. He climaxed his paper with the following sweeping generalizations:

"There are no doubt sufficiently profound distinctions between the ethical process and the cosmic process as it existed prior to man and to the formation of human society. So far as I know, however, all of these differences are summed up in the fact that the process and the forces bound up with the cosmic have come to consciousness in man. That which was instinct in the animal is conscious impulse in man. That which was 'tendency to vary' in the animal is conscious foresight in man. That which was unconscious adaptation and survival in the animal, taking place by the 'cut and try' method until it worked itself out, is with man conscious deliberation and experimentation. That this transfer from unconsciousness to consciousness has immense importance, need hardly be argued. It is enough to say that it means the whole distinction of the moral from the unmoral."[2]

Long ago, Thomas Henry Huxley, the great evolutionary propagandist, clearly recognized that evolutionary history implied a basic principle in nature of aggressive self-interest. The great evolutionary slogans of the time — "struggle for existence" and "survival of the fittest" — clearly suggested a naturalistic ethic that continually pitted individuals, races, species, classes, and nations against each other. He also saw that this system squarely contradicted the Christian ethic, and made this the

---

[1] H. J. Muller: *op cit.*, p. 628.
[2] John Dewey: "Evolution and Ethics," reprinted in *Scientific Monthly*, February 1954, p. 66. Originally published in *The Monist*, Vol. VIII. (1897-1901).

theme of his Romanes lectures, *Evolution and Ethics*. In the present century, one of the leading evolutionists of the interwar period, Sir Arthur Keith, published another influential book with the same title. The entire volume amplifies and reinforces the theme of Huxley; namely, that the ethics taught by Christ and by evolution are polar opposites. Among other things, he says:

"If the final purpose of our existence is that which has been and is being worked out under the discipline of evolutionary law, then, although we are quite unconscious of the end result, we ought, as Dr. Waddington has urged, to help on 'that which tends to promote the ultimate course of evolution.' If we do so, then we have to abandon the hope of ever attaining a universal system of ethics; for, as we have just seen, the ways of national evolution, both in the past and in the present, are cruel, brutal, ruthless, and without mercy." [1]

Sir Arthur makes an interesting observation in the following:

"It was often said in 1914 that Darwin's doctrine of evolution had bred war in Europe, particularly in Germany. An expression of this belief is still to be met with. In 1935, a committee of psychologists, representing thirty nations, issued a manifesto in which it was stated that 'war is the necessary outcome of Darwin's theory . . .' The law of evolution, as formulated by Darwin, provides an explanation of wars between nations, the only reasonable explanation known to me." [2]

Of course, the Christian would say, rather, that wars result from sin, and that the only hope for permanent peace is in Jesus Christ. But it is clear that the inexorable logic of evolutionary reasoning leads directly to the conclusion that war and struggle is the chief good, leading to evolutionary advance.

"Meantime, let me say that the conclusion I have come to is this: the law of Christ is incompatible with the law of evolution — as far as the law of evolution has worked hitherto. Nay, the two laws are at war with each other; the law of Christ can never prevail until the law of evolution is destroyed." [3]

---

[1] Sir Arthur Keith: *Evolution and Ethics* (New York, Putnam, 1947), p. 15.
[2] *Ibid*, p. 149.
[3] *Ibid*, p. 15.

Now, admittedly, such views are not shared by most modern evolutionists, especially in America, where war has suddenly become so unpopular. Nevertheless, Sir Arthur is recognized by all to be one of the greatest evolutionists of modern times, and he had given more study to this subject of evolutionary ethics than any of his contemporaries. Many modern evolutionists like to stress "cooperation" as a viable force in evolutionary advance, but they do so without much conviction. Even such a cautious scientist as Dr. Frederick Seitz, when serving as President of the National Academy of Sciences, has said:

"We can, of course, be grateful to nature for the highly remarkable genetic gifts which we have inherited as a result of the very complex process of selection which our ancestors experienced. We must also keep in mind, however, that many of our most valued characteristics probably emerged out of interhuman competition. We probably have instinctive patterns of behavior which are fundamentally inimical to human cooperation on an indefinitely broad scale."[1]

Religion

Not even the domain of religion has escaped the influence of evolution. We should remember the classic diatribe of Sir Julian Huxley, who could probably be justly identified as the world's number-one modern evolutionist. As the keynote speaker at the great Darwinian Centennial Convocation in 1959 at the University of Chicago, he orated as follows:

"In the evolutionary system of thought there is no longer need or room for the supernatural. The earth was not created; it evolved. So did all the animals and plants that inhabit it, including our human selves, mind and soul, as well as brain and body. So did religion. Evolutionary man can no longer take refuge from his loneliness by creeping for shelter into the arms of a divinized father figure whom he himself has created."[2]

Of course, with the weight of all the leading scientific authorities on the side of evolution, religious leaders around the

---

[1] Frederick Seitz: "Science and Modern Man," *American Scientist,* Vol. 54, September 1966, p. 230.
[2] Sir Julian Huxley: *Associated Press* dispatch, November 27, 1959.

world have felt it necessary to devise systems for accommodating their systems within the evolutionary framework. This has been a relatively easy adjustment for most of the non-Christian religions, which were all fundamentally evolutionary systems anyway.

Buddhism and Hinduism, with their doctrines of *karma* and their pantheistic conceptions of God; Confucianism and Taoism, with their essentially agnostic attitude toward the idea of a personal God; and Shintoism, with its deification of man and the state, are all fundamentally evolutionist in philosophy and have quickly and easily adapted themselves to the Darwinian approach within the framework of their own systems.

The same is true of the animistic faiths, whenever their practitioners become sufficiently sophisticated in their understanding of the modern world through education (often, sad to say, provided by "Christian" nations, and even sometimes in missionary schools). Fundamentalist missionaries in Africa and other areas of tribal religions report that the teaching of evolution in the schools, together with its adaptation as a veneer over-revived demon worship, is today one of the most serious hindrances to the Gospel in such lands.

In their primitive form (or, better, degenerate forms, as derived from primitive monotheism in the manner described in the Bible in Romans 1:18-32), these animistic religions are themselves crudely evolutionist, all believing in some form of magical development of the first men and animals from previous materials. Though they retain in some cases a very faint and impersonal tribal memory of a "high God" of some sort, their practical daily religion has altogether to do with the physical world, and its control by the spirits of departed ancestors and demons.

Modern spiritism is essentially the same thing, and this religion, together with its varied, associated cults — astrology, witchcraft, Satanism, theosophy, Zen Buddhism, and the like — has been in recent years sweeping like wildfire over the world, especially Europe and Latin America. In spite of their individual differences, all such occult religions uniformly make the claim that they are more in accord with the modern scientific evolutionary view of the world than are the traditional religions, especially Christianity.

Even the various pseudo-Christian cults, such as Christian Science, Unity, and others of like kind, have commonly accepted evolution into their systems. In somewhat analogous fashion the parabiblical religions of Judaism and Islam, though nominally committed to faith in the Genesis account of creation, have now largely capitulated to the evolutionary cosmology in their philosophies. Reformed Judaism and Conservative Judaism have almost completely accepted Darwinian evolution and the higher critical views of the Old Testament and even many Orthodox Jews have adopted a symbolic interpretation of the Genesis record.

The Muslim religious leaders have largely done the same, although there are minorities who adhere to the literal creation of all things in the beginning by Allah. The Muslim mystics and philosophers, with their wholly transcendent view of God's nature and their subjective approach to religious experience, have in fact to all practical purposes been evolutionary pantheists all along.

In general, it is realistic to say that practically the whole world of religion — Christian, non-Christian, and quasi-Christian alike — has accepted the evolutionary cosmology in one form or another.

Christianity

Some uninformed Christians might object that, surely, Christian people still reject evolution. However, this is not true at all, at least for the great majority of those who at least profess to be Christians.

The history of organized Christianity in the past century has been in large measure a sad record of compromise and retreat before the attacks of the evolutionists. In the nineteenth century, the so-called "higher criticism" launched its attacks against the authenticity of the Old Testament, especially the books of Moses, leading to the "documentary hypothesis" of the composition of the Old Testament writings. Many conservative seminaries and Bible schools today devote much study to the refutation of this higher criticism, but ignore the evolutionary philosophy which spawned it.

"Sometimes people talk as though the 'higher criticism' of texts in recent times has had more influence than the higher criticism of nature. This seems to me to be nonsense. The higher criticism has been simply an application of an awakened critical faculty to a particular kind of material, and was encouraged by the achievement of this faculty to form its bold conclusions. If the biologists, the geologists, the astronomers, the anthropologists, had not been at work, I venture to think that the higher critics would have been either non-existent or a tiny minority in a world of fundamentalists."[1]

Even Bible-believing Christians who have rejected the higher critical views of Genesis, have again and again tried to compromise with evolution by novel exegetical twists of Genesis. This is always only a superficial and temporary stratagem, and inevitably culminates in a reinterpretation of the entire Christian faith to correlate with the full-orbed evolutionary view of the world and society. Today, almost all of the colleges and seminaries of the large denominations, even those that are supposed to be conservative, have incorporated the evolutionary system and its associated liberal or neo-orthodox theology into their teachings. Even many schools which until recent years were strongly fundamental and anti-evolutionary are now allowing theistic evolution (or its semantic substitute, progressive creation) as a legitimate position for their faculty to teach if they so choose. This compromise has already in many cases been followed by a weakened doctrine of Biblical inspiration, and increased conformity to the world in areas of morals and social activism.

Totalitarian Ideologies

A survey of the present status of evolutionary thought would be seriously incomplete without a consideration of its determinative influence on Communism, Nazism, and similar collectivist systems. These systems are actually religions, believed and promulgated through the fanatical faith of their devotees.

---

[1] F. M. Powicke: *Modern Historians and the Study of History, Essays* (London, 1955), p. 228.

Though not much in vogue currently, the fascistic systems of Hitler, Mussolini, and others almost conquered the world a generation ago. There are even now neo-Nazi movements which bear watching, as well as various dictatorships of similar character around the world; not to mention the "new left" student movement which strangely resembled the early days of Nazism.

In any case, all such ideologies, built up as they are on the concepts of racism and statist totalitarian aggression and control, are direct products of the Darwinian doctrines of struggle for existence and survival of the fittest. Friedrich Nietzsche, the philosophical father of these systems, was an ardent evolutionist, as were his spiritual children, Hitler and Mussolini.

"From the 'Preservation of Favoured Races in the Struggle for Life' (i.e., Darwin's subtitle to *Origin of Species*) it was a short step to the preservation of favored individuals, classes or nations — and from their preservation to their glorification. Social Darwinism has often been understood in this sense: as a philosophy, exalting competition, power, and violence over convention, ethics, and religion. Thus it has become a portmanteau of nationalism, imperialism, militarism and dictatorship, of the cults of the hero, the superman, and the master race . . . recent expressions of this philosophy, such as (Hitler's) *Mein Kampf*, are, unhappily, too familiar to require exposition here. And it is by an obvious process of analogy and deduction that they are said to derive from Darwinism. . . . Nietzsche predicted that this would be the consequence if the Darwinian theory gained general acceptance." [1]

Fascism is generally held to be a right-wing movement and Communism a left-wing. Both, however, are variants of the same species, evolutionistic, totalitarian collectivism. In any case, it is clear that the basic rationale of Communism, just as that of Fascism, is the dogma of materialistic evolution.

"In an age of social Darwinism, the combination of the ideas of struggle, of historical evolution, and of progress proved

1 Gertrude Himmelfarb: *Darwin and the Darwinian Revolution* (London, Chatto and Windus Publ., 1959), pp. 343-44.

irresistible. The Marxists became merely a sect in the larger church . . . . In urging these lessons, Marx and Engels set the pattern of all subsequent Marxist polemics by using what may be called the evolutionist's double standard: when *you* do it, it's wrong, because you are the past; when *we* do it, it's right, for we are the future. The mood — borrowed from science — is that of a mighty ruthlessness. History, like nature, is tough." [1]

Conway Zirkle, of Pennsylvania State University, discusses at considerable length the close connection between Darwinism and Marxism.

"Evolution, of course, was just what the founders of communism needed to explain how mankind could have come into being without the intervention of any supernatural force, and consequently it could be used to bolster the foundations of their materialistic philosophy."[2]

It is well known that not only the early Communists, such as Marx and Engels, were atheistic evolutionists, but also that all the leaders of Communism since have been the same. Though they have fluctuated between Darwinian and Lamarckian biology in their application of evolutionary mechanisms to communist theory, they have never varied in their commitment to evolution itself.

Other applications of social Darwinism could be discussed, such as the dog-eat-dog capitalism and the military imperialism of the nineteenth century. It seems that every ideology which has majored in struggle for supremacy of race, class, or other social group over others has been structured around evolution.

"In turn, biological evolutionism exerted ever-widening influences on the natural and social sciences, as well as on philosophy and even on politics. Not all of these extra-biological repercussions were either sound or commendable. Suffice it to mention the so-called Social Darwinism, which often sought to justify the inhumanity of man to man, and the biological racism which furnished a fraudulent scientific sanction for the

---

[1] Jacques Barzun: *Darwin, Marx, Wagner* (New York, Doubleday, 1958), p. 186.

[2] Conway Zirkle: *Evolution, Biology and the Social Scene,* (Philadelphia, University of Pennsylvania Press, 1959), p. 85.

atrocities committed in Hitler's Germany and elsewhere."[1]

It might be appropriate to refer again at this point to the classic volume on evolutionary ethics by Sir Arthur Keith. Modern evolutionists have sometimes denounced the works of Barzun and Himmelfarb, quoted above, despite their full documentation, on the basis that neither was a scientist. However, all evolutionists would surely recognize Sir Arthur as one of their own.

When he wrote his book, he had just been through World War II, enduring with other Britons the awful suffering visited by Adolph Hitler on England and the world. He certainly did not write out of any feeling of sympathy for Hitler and his cause. Yet his honest understanding of the real nature of evolution, in which he firmly and fully believed, impelled him to say:

"To see evolutionary measures and tribal morality being applied vigorously to the affairs of a great modern nation, we must turn again to Germany of 1942. We see Hitler devoutly convinced that evolution produces the only real basis for a national policy. . . . The means he adopted to secure the destiny of his race and people were organized slaughter, which has drenched Europe in blood . . . . Such conduct is highly immoral as measured by every scale of ethics, yet Germany justifies it; it is consonant with tribal or evolutionary morality. Germany has reverted to the tribal past, and is demonstrating to the world, in their naked ferocity, the methods of evolution."[2]

And again, he says:

"The German Fuhrer, as I have consistently maintained, is an evolutionist; he has consciously sought to make the practice of Germany conform to the theory of evolution. He has failed, not because the theory of evolution is false, but because he has made three fatal blunders in its application."[3]

Modern American evolutionists may be embarrassed by this philosophical association with the Fascism and Nazism of Mussolini and Hitler, but it is nevertheless a fact, and it is a fact which certainly ought to awaken *theistic* evolutionists at least to the real nature of the theory with which they have been willing to

---

1 Theodosius Dobzhansky: "Evolution at Work," *Science,* May 9, 1958, p. 1091.
2 Sir Arthur Keith: *Evolution and Ethics,* (New York, Putnam, 1947), p. 28.
3 *Ibid,* p. 230.

become identified for the sake of academic prestige. If one really feels he must believe in evolution, he should at least leave God out of it. The very idea of "theistic" evolution is a contradiction in terms, about like "theistic atheism" or "flaming snowflakes." The evil fruits of evolution are strong evidence of its bitter roots.

Racism

While discussing totalitarian philosophies, it is appropriate to point out that racism also, in the form of racial hatreds, racial warfare, the assumption of racial superiority or inferiority, and other such virulent offshoots, is strictly a product of evolutionary thinking. The Bible does not use either the word or the concept of "race" anywhere at all. In God's created economy, there is no such thing as a "race," but only nations and tribes.

This is not to say, of course, that modern evolutionists are all racists. It is no longer popular in America to be a racist, and liberals of every academic persuasion, including most (not all) modern evolutionary biologists decry and eschew racism.

This was not true of the nineteenth century evolutionists, however. A recent book[1] has thoroughly reviewed this subject and demonstrated that literally *all* of the nineteenth century evolutionists believed in the evolutionary superiority of the white race and the inferiority of the others, especially the Negro. A recent review says:

> "This is an extremely important book, documenting as it does what has long been suspected: the ingrained, firm, and almost unanimous racism of North American men of science during the nineteenth (and into the twentieth) century. . . . *Ab initio,* Afro-Americans were viewed by these intellectuals as being in certain ways unredeemably, unchangeably, irrevocably inferior."[2]

A reviewer in another scientific journal notes the following from the book:

> "What was new in the Victorian period was Darwinism. . . .

---

1 John S. Haller, Jr.: *Outcasts from Evolution: Scientific Attitudes of Racial Inferiority, 1859-1900* (Urbana, University of Illinois Press, 1971), 228 pages.

2 Sidney M. Mintz: *American Scientist,* Vol. 60, May-June 1972, p. 387.

Before 1859, many scientists had questioned whether blacks were of the same species as whites. After 1859, the evolutionary schema raised additional questions, particularly whether or not Afro-Americans could survive competition with their white near-relations. The momentous answer was a resounding no. . . .
was inferior because he represented the 'missing
ape and Teuton.' "[1]

d of evolutionary thinking essentially universal, it
hat the concepts of race were so important in the
f the master-race idea. Not only, however, were
rs of his stripe ardent evolutionary racists, but so
'x and his socialist and communist colleagues, all
evolutionary considerations.
this remark, the present writer asked the ad-
whether this doctrine would not imply that the
iority, and primitive peoples, those who had had
or mental and physical development, were not also
ess advanced than the dominant ones. 'Ah yes,' he
nfidential manner and after some hesitation, 'yes,
nit that this is, after all, true. They are in fact in-
iologically in every respect, including their heredi-
' he added, 'is in fact the official doctrine.' "[2]

tter, so was Charles Darwin himself. Although he
o slavery, he too thought that the Negro was of a
e, and was doomed to become extinct in future
ompetition with the more favored races.[3]
a of "race," of course, is an evolutionary concept,
or theological concept. A race is essentially a sub-
l, if isolated long enough, may well evolve into a
Although most modern evolutionists hold to the
(single line) rather than polyphyletic origin of man,
'e that the various races have been distinct for at
score thousand years, so that ample time has been
volve significant differences between them. This

The third son of King Juao, Henry used his interest in geography and seamanship to found a school to promote the study of navigation. He supported the exploration of the African coast and developed caravel ships which were used in exploration for the next century.

am: *Science,* Vol. 175, February 4, 1972, p. 506.
: *op cit.,* p. 335.
II, p. 164.

possibility is mentioned even by such an orthodox evolutionist as George Gaylord Simpson, who would of course indignantly deny any charge of racism against himself.

"Evolution does not necessarily proceed at the same rate in different populations, so that among many groups of animals it is possible to find some species that have evolved more slowly, hence are now more primitive, as regards some particular trait or even over-all. It is natural to ask — as many have asked — whether among human races there may not similarly be some that are more primitive in one way or another or in general. It is indeed possible to find single characteristics that are probably more advanced or more primitive in one race than in another."[1]

Now of course our whole point in this discussion is to show that, rightly or wrongly, evolutionary thinking is at the root of modern racism and racial conflicts. Once again, this is not meant at all to imply that all or most modern evolutionists are themselves racists. It is just that evolution itself is fundamentally racist. As Keith says:

"Christianity makes no distinction of race or of color; it seeks to break down all racial barriers. In this respect the hand of Christianity is against that of Nature, for are not the races of mankind the evolutionary harvest which Nature has toiled through long ages to produce? May we not say, then, that Christianity is anti-evolutionary in its aim? This may be a merit, but if so it is one which has not been openly acknowledged by Christian philosophers."[2]

Control of Future Evolution

One of the frightening things about modern evolutionists and sociologists is that they have come to believe they should control future evolution. This they propose to do by genetic manipulations of various sorts, by control of births and possibly of deaths, by an intellectual elite who will decide who is fit to have children and what kinds of babies are desirable, and by state

---

[1] George Gaylord Simpson: "The Biological Nature of Man," *Science,* Vol. 152, April 22, 1966, p. 475.

[2] Sir Arthur Keith: *Evolution and Ethics,* (New York, Putnam, 1947), p. 72.

enforcement of their decisions. The dangers of the population explosion and nuclear war are believed to justify such future control of evolutionary planning.

One of the leading advocates of controlled evolution has been H. J. Muller, admittedly one of the world's top geneticists and most influential leaders of evolutionary thought. In one of the many papers he has published on this subject, he says:

"It is indisputable that, as man's control over matter advances, more and more of his bodily structure and functioning can be amended to advantage, and even replaced by artificial means . . . the time-honored notions of how reproduction should be managed will gradually give way before the technological progress that is opening up new and ever more promising possibilities."[1]

In an earlier paper, Muller had said:

"It has rightly been said that biological evolution is multi-directional and cruel and that the vast majority of lines of descent end in pitiful anticlimaxes."[2]

This, of course, is exactly the reason why we are sure the God of the Bible could have had nothing to do with such a process. However, instead of acknowledging that it is the evolutionary idea itself which is wrong, Muller proposes to correct the methods of evolution and make them more efficient:

"Through the unprecedented faculty of long-range foresight, jointly serviced and exercised by us, we can, in securing and advancing our position, increasingly avoid the missteps of blind nature, circumvent its cruelties, reform our own natures, and enhance our own values."[3]

That does sound like quite an ambitious program for mortal man! But many others are convinced man can do it, among them Hudson Hoagland:

"Man's unique characteristic among animals is his ability to

---

[1] H. J. Muller: "Should We Weaken or Strengthen Our Genetic Heritage?" in *Evolution and Man's Progress* (Ed. by Hudson Hoagland and Ralph Burhoe, New York, Columbia University Press, 1962), pp. 24, 32.

[2] H. J. Muller: "Human Values in Relation to Evolution," *Science,* Vol. 127, March 21, 1958, p. 629.

[3] *Ibid.*

direct and control his own evolution, and science is his most powerful tool for doing this."[1]

More recently, Dr. A. G. Motulsky, Director of the Center for Inherited Diseases at the University of Washington, has confirmed this opinion.

"We no longer need be subject to blind external forces but can manipulate the environment and eventually may be able to manipulate our genes. Thus, unlike any other species, we may be able to interfere with our biological evolution."[2]

Thus, man has not only eliminated God from history but now proposes to play God himself.

[1] Hudson Hoagland: "Science and the New Humanism," *Science,* Vol. 143, January 10, 1964, p. 111.

[2] A. G. Motulsky: "Brave New World?" *Science,* Vol. 185, August 23, 1974, p. 653.

Modern humanists say that man has gradually evolved from primeval chaotic stuff, by a strange trial-and-error process of evolutionary meandering over billions of years. The fact is that the evolutionary theory itself has evolved from the cosmic idolatries of the ancient pagans to the naturalistic speculations of the modern Darwinists, but it has always been the same old muddy stream of man-centered religion.

San Juan River, Colorado

Photo credit: Dave Kerr

CHAPTER III

A LONG, LONG
TRAIL A'WINDING

The Darwinian Century

In the previous chapter we have traced the pervasive influence of the evolutionary philosophy as it has affected almost every discipline and segment of modern culture, especially in the educational system. Obviously such profound and widespread effects must have an adequate background cause. This did not all happen by chance nor did it develop overnight. There has been a long history of evolutionary thought and there have been powerful forces at work to bring about its universal acceptance. Especially must this be true in view of the fact that both the inspired Word of God and all the real data of science support creation rather than evolution. In this chapter, therefore, we shall in outline fashion attempt to trace some of this history, especially the pre-Darwinian history, of evolution.

The theory of evolution as held today is of course associated most commonly with Charles Darwin. It is doubtful if any other scientist has ever received as much praise and adulation as has Darwin. He was even honored in what amounted almost to a religious worship service in a great convocation held in 1959 at the University of Chicago, on the occasion of the one-hundredth anniversary of the publication of his famous book *The Origin of Species by Natural Selection.*

And yet the actual scientific accomplishments of Charles Darwin were relatively insignificant, at least by modern standards. The only college degree he ever earned was in theology, and this with a very undistinguished record. He was not inclined toward a career as a clergyman, and was in fact only a nominal believer in Christianity, at most. He became a botanist more or less by accident, spending five years on an extended voyage around the world as a ship's naturalist on the *Beagle.* This voyage began in 1831, when he was only 22 years old, and lasted five years. It was at this time he read Lyell's *Principles of Geology* which, added to his previous reading of Thomas Malthus' *Principles of Population,* and his own observations on the fauna of the South Pacific islands, quickly led to his full acceptance of evolution and his postulation of natural selection as its mechanism.

He also published studies on barnacles, vegetable mold and other more restricted biological subjects, but these were hardly of great significance. It was clearly his *Origin of Species* which acquired for him his super-reputation.

This situation is odd, to say the least. Charles Darwin certainly was not the inventor of the theory of evolution, as many suppose.

"The idea of evolution had been widespread for more than 100 years before 1859. Evolutionary interpretations were advanced increasingly often in the second half of the 18th and the first half of the 19th centuries." [1]

Neither did he originate the idea of natural selection, although this was his frequent boast. As a matter of fact, he rushed his book into print before he was ready (although he had been working on it well over 20 years), in fear that Alfred Wallace would publish his own identical theory first.

---

[1] Ernst Mayr: "The Nature of the Darwinian Revolution," *Science,* Vol. 176, June 2, 1972, p. 981.

"Wallace independently achieved and set forth the same ideas as Darwin. He was an independent discoverer of natural selection."[1]

Nevertheless, his famous book *The Origin of Species by Natural Selection* was published in 1859 and immediately created a storm of controversy which has continued for over a hundred years. It is an interesting comment on the temper of the times and man's eagerness to discover a justifiable reason for rejecting God as his Creator that the first edition of the *Origin* was sold out before it was published.

Despite considerable opposition at first, Darwin's theories were soon largely accepted, both by scientists and religionists. The post-Darwinian century has beyond doubt, almost from the first, been one in which evolutionary thought has reigned supreme. There have been so many volumes written on this subject that it is essentially common knowledge.

The most significant fact about Darwin is not his stature as a scientist but his influence as a symbol. His contribution came at just the right time to catalyze an explosive reaction, transforming in one generation a society which was already seething in inner rebellion against the predominant theological and Biblical view of the world into a world in open and often violent rebellion against its Creator.

Darwin has often been called the "Newton of biology" but Jacques Barzun shows that this is a "very loose description indeed." Barzun (Professor of History and Dean of the Graduate Faculties at Columbia University) says:

"Darwin was not a thinker and he did not originate the ideas that he used. He vacillated, added, retracted, and confused his own traces. As soon as he crossed the dividing line between the realm of events and the realm of theory he became 'metaphysical' in the bad sense. His power of drawing out the implications of his theories was at no time very remarkable, but when it came to the moral order it disappeared altogether,

---

[1] Loren C. Eiseley: "Alfred Russell Wallace," *Scientific American,* Vol. 199, p. 70.

as that penetrating evolutionist, Nietzsche, observed with some disdain.''[1]

Nevertheless, despite his serious deficiencies as a scientist, thinker and writer, as Barzun also says:

"Clearly, both believers and unbelievers in Natural Selection agreed that Darwinism had succeeded as an orthodoxy, as a rallying point for innumerable scientific, philosophical, and social movements. Darwin had been the oracle and the *Origin of Species* the 'fixed point with which evolution moved the world.' "[2]

Therefore, as we try to trace the history of the development and influence of evolutionary thought, it is evident that neither Charles Darwin nor his famous book provides the answer. We must look deeper and farther than Darwin to find the real roots of evolutionism and its worldwide influence.

Why Darwin?

Although the name of Charles Darwin is honored — in fact almost revered by many — as the founder of the modern theory of evolution, it is largely an artificial and manufactured identification. His name serves a purpose as a sort of symbol and rallying point for evolutionists, but his actual scientific accomplishments are rather ordinary and unimpressive by modern standards. Although many intellectuals seem to feel it is sort of a union card to pay due homage to him, one senses in their writings a feeling that they are not comfortable in this nor quite clear what it is that they admire in Darwin and his work. There seems to be something in these developments that has never yet been satisfactorily understood or explained.

The theory of evolution did not originate with Charles Darwin, of course, as noted above. Various evolutionary concepts were accepted by many people long before Darwin. Darwin's main contribution was the theory of natural selection as an evolutionary mechanism, but even this was not original with him. In an authoritative review of the origins of Darwinism, an outstanding British biologist says:

---

1 Jacques Barzun: *Darwin, Marx, Wagner* (2nd Ed., Garden City, N.Y., Doubleday and Co., 1959), p. 84.

2 *Ibid*, p. 69.

"As to the means of transformation, however, Erasmus Darwin originated almost every idea that has since appeared in evolutionary theory. . . . All three men (i.e., William C. Wells, James C. Pritchard, and William Laurence) advanced explicitly and in detail the alternative theory of natural selection foreshadowed by Erasmus Darwin. These three communications were the first clear statements of such an idea — in opposition to notions of evolution guided by design and purpose — since classical times."[1]

Erasmus Darwin was Charles Darwin's grandfather and was a widely-read and popular writer on evolution even before Charles was born. Wells, Pritchard, and Laurence were all physicians who wrote on evolution and natural selection almost a half-century before *The Origin of Species.* Diderot in France, Edward Blyth in England, and even Benjamin Franklin advanced similar theories. However, Charles Darwin never acknowledged his predecessors and always called natural selection "my theory."

He was also much influenced by other evolutionary predecessors such as Lamarck, who was most famous for his theory of evolution by acquired characters, and Chambers, who had written a very influential work advocating theistic evolution called *Vestiges of the Natural History of Creation.* Although Darwin denounced Lamarck's theory, he later superimposed much of it on his own theory.

Darwin was also much influenced by Thomas Malthus and his concept of the "struggle for existence" among human populations. But probably his most important immediate predecessor was Charles Lyell, with his geological dogma of uniformitarianism. The vast span of geologic time, with its supposed gradual progression of life forms into systems of higher and higher complexity, was of course an absolute necessity for any viable theory of evolution. Darwin frequently acknowledged his debt to Lyell, even though he was evidently always quite reluctant to give credit to his other predecessors. Lyell rejected the predominant catastrophist theory of geology and persuaded

---

[1] C. D. Darlington: "The Origin of Darwinism," *Scientific American,* Vol. 201, May 1959, p. 62.

his contemporaries that all the geologic strata had been laid down slowly over vast ages of time. Darwin found this framework made to order for his ideas of natural selection, which would certainly require tremendous lengths of time to be effective.

"Darwin and Wallace were Lyell's intellectual children. Both would have failed to be what they were without the *Principles of Geology* to guide them."[1]

With respect to the so-called "evidences for evolution," these had been well expounded long before 1859. Darlington notes:

"In favor of the evolution of animals from 'one living filament.' Erasmus Darwin (who died before Charles was born — author) assembled the evidence of embryology, comparative anatomy, systematics, geographical distributions and, so far as man is concerned, the facts of history and medicine. . . . These arguments about the fact of transformation were all of them already familiar. As to the means of transformation, however, Erasmus Darwin originated almost every important idea that has ever appeared in evolutionary theory."[2]

It does seem strange that Charles Darwin would never acknowledge his intellectual debt to these predecessors. His admirers speak of him as though he were the paragon of the careful, open-minded scientist, humble, interested solely in the hard-headed observation and understanding of facts. One of these admirers, George Gaylord Simpson, in a review of his autobiography, commented thus:

"Darwin himself . . . wrote he 'never happened to come across a single naturalist who seemed to doubt about the permanence of species,' and he acknowledged no debt to his predecessors. These are extraordinary statements. They cannot be literally true, yet Darwin cannot be consciously lying, and he therefore may be judged unconsciously misleading, naive, forgetful, or all three. His own grandfather, Erasmus Darwin, whose work Charles knew very well, was a pioneer evolutionist. Darwin was also familiar with the work of Lamarck, and had certainly

---

[1] Loren C. Eisely: "Charles Lyell," *Scientific American,* Vol. 201, August 1959, p. 106.

[2] C. D. Darlington: "The Origin of Darwinism," *Scientific American,* Vol. 201, May 1959, pp. 61, 62.

met a few naturalists who had flirted with the idea of evolution
. . . . Of all this, Darwin says that none of these forerunners had
any effect on him. Then, in almost the next breath, he admits
that hearing evolutionary views supported and praised rather
early in life may have favored his upholding them later."[1]

Simpson tries to excuse him in the same vein as the writer of
his biography, Nora Barlow, his granddaughter, who said that he
had simply dismissed all these predecessors because they did not
back up their theories with evidence. Of course, when it comes to
real *evidence*, Charles Darwin had none either. He may have
found examples in which natural selection was effective in
weeding out unfit varieties, but none in which natural selection
produced new characteristics which were favorable and yet
could not have resulted from normal variations within the kind.

The most illustrious of Darwin's evolutionist predecessors was
probably Lamarck. He had of course proposed as his evolutionary
mechanism the inheritance of acquired characteristics, and had
written in great depth upon the process and meaning of evolution.
Darwin bitterly attacked Lamarck's ideas, yet later he gradually
incorporated many of them into his own system, including a
modified form of acquired character inheritance. Darlington
says:

"Darwin was slippery . . . (using) a flexible strategy which is
not to be reconciled with even average intellectual integrity. . . .
He began more and more to grudge praise to those who had in
fact paved the way for him. . . . Darwin damned Lamarck and
also his grandfather for being very ill-dressed fellows at the
same moment he was engaged in stealing their clothes."[2]

There was also Herbert Spencer, whose writings had a profound
influence on the acceptance of evolution in the 19th century. It
was Spencer who coined the Darwinian phrases "struggle for existence" and "survival of the fittest," and he was writing on
evolution for some time before Darwin began to do so.

---

[1] George G. Simpson: "Charles Darwin in Search of Himself," Review of
Autobiography of Charles Darwin, by Nora Barlow, *Scientific American,*
Vol. 200, August 1959, p. 119.
[2] C. D. Darlington: *Darwin's Place in History* (London, 1959), pp. 60, 62.

". . . in his own day, which was that of Darwin, too, Spencer was regarded as a giant, and his *Principles of Biology* was adduced as one of the chief evidences for this high estimation. Of course, this could not be on literary grounds. Spencer is no more a first-class stylist than Darwin. . . . Had Darwin and Spencer been more tendentious men, they would doubtless have become embroiled in Newton-Leibniz disputes regarding priorities. . . . It would be difficult to establish the interlocking priorities here. Spencer's preliminary essays were published some time before *The Origin of Species.*"[1]

We should also of course mention Thomas Huxley, who was probably more responsible than any other single individual for the rapid and widespread acceptance of Darwinian evolution, through his constant and effective speaking and writing. Huxley was an evolutionist before Darwin, but the latter's book gave him the needed scientific support for it, or so he thought. He became known as "Darwin's bulldog." He opposed not only creationism, but Christianity in general, lecturing against the resurrection of Christ and other Biblical truths, and applying to himself the term *agnostic,* which he invented. That he was not without ulterior motives in all this is evident in the fact that he had a driving ambition for fame and historical recognition:

"He had a work to do in England, a messianic purpose, and he dedicated to that purpose his tireless energy and his vast resources of knowledge and ability. And he did attain the success his heart desired, for Huxley was recognized as a prophet in his own country."[2]

Thus in no sense could Charles Darwin be said to be responsible for the theory of evolution. The remarkable chain of events leading up to the publication of his book, and the even more remarkable results of its publication, have not even yet been satisfactorily explained. One thing is sure — it was not simply the

---

[1] George Kimball Plochmann: "Darwin or Spencer?" *Science,* Vol. 130, November 27, 1959, p. 1452.
[2] Charles S. Blinderman: "Thomas Henry Huxley," *Scientific Monthly,* April 1957, p. 172.

instant triumph of science over superstition that evolutionary propagandists like to suppose.[1]

These developments were not taking place in a vacuum. Most scientific writers on the history of evolution write as though the whole story is simply one of the advance of science, and the delivery of the people from the cloud of ignorance and superstition that had covered them before Darwin. But the middle of the 19th century was a time of great political and social ferment, and the scientists as well as others were being caught up in these movements. Europe was still involved in the aftermath of the French Revolution, and other revolutionary movements were seething everywhere. Socialism and communism were familiar terms and so was anarchism. These were the times of Hegel and Karl Marx, of the industrial revolution and the American Civil War, of Nietzsche and the growth of economic imperialism.

It has become almost a cliche, oft-repeated, that "evolution was in the air" and, for that matter, so was "revolution." All the great political and economic movements — whether communism, economic imperialism, centralized capitalism, racism, or others — were all eager to grab up Darwinism as scientific justification for their particular brands of man's basic self-centered struggle-and-survival ethic.

Although faith in the Bible and creation was still very strong in both Europe and America, resulting from the spiritual revivals of the Reformation Period and of the great awakening, there had been strong undercurrents of unbelief for a long time. Subversive revolutionary movements were influencing multitudes. Deist philosophers, Unitarian theologians, Illuminist conspirators, Masonic syncretists, and others were all exerting strong influences away from Biblical Christianity and back to paganistic pantheism. The French Revolution had injected its poisons of atheism and immoralism into Europe's bloodstream, and the German rationalistic philosophers had laid the groundwork for

---

[1] Darlington says: "He was able to put his ideas across not so much because of his scientific integrity, but because of his opportunism, his equivocation and his lack of historical sense. Though his admirers will not like to believe it, he accomplished his revolution by personal weakness and strategic talent more than by scientific virtue." ("The Origin of Darwinism," p. 66.)

the destruction of Biblical theology in the schools and churches. Socialism and communism were on the upswing throughout Europe; Marx and Nietzsche were propagating their deadly theories and were acquiring many disciples — perhaps also financial backers, as students of conspiracies have frequently suggested. All of these people and movements were evolutionists of one breed or another.

It is significant that all the above movements and many others of like kind had rejected the Biblical cosmology and followed an evolutionary cosmology as their basic rationale. However, there were two strong barriers holding back the tide of paganism. In the first place, the Christian churches and schools had been strengthened by the works of many great Christian apologists (e.g., Paley, Lyttelton, West, Butler, Edwards, Dwight, *et al*), whose labors had all but demolished the systems of Deism and Unitarianism. In the second place, the industrial revolution had drastically increased public awareness and respect for science and technology, and the great scientists of the day were mostly Bible-believing Christians. Scientific philosophy was largely structured around natural theology, and scientific study was considered to be, as the great Isaac Newton had said, "thinking God's thoughts after Him." All the data of natural science were understood as supporting the fact of divine creation and providence; even the sediments and fossils of the new science of geology were understood in terms of the great Flood.

Sir Isaac Newton is generally acknowledged to have had the greatest scientific intellect of all time, and the weight of his great authority had long been cited in favor of belief in the full authority of the Bible. He, as did his successor at Cambridge, John Woodward, believed in the literal creation account of Genesis, as well as that of the worldwide Flood. In fact, Newton was a colleague and follower of the great scholar, Archbishop Ussher, whose Biblical chronology is unjustly ridiculed today by multitudes who exhibit only a small fraction of the ability and careful scholarship which Ussher manifested in his day.

In fact, the period from 1650 to 1850 was an era of scientific giants, and many of the greatest among them — men such as Pascal, Faraday, Maxwell, Kelvin, and others — were men who believed in the inspiration and authority of the Bible. The marvelous discoveries and achievements of science, revealing the complexities and orderly relationships in nature, seemed more and more to confirm the fact of design and therefore the existence of a Creator.

Therefore, if the great complex of anti-Christian movements and philosophies was to be successful in its struggle for control of the minds and hearts of men, something would have to be done first of all to undermine Biblical creation and to establish evolution as the accepted cosmogony. The Biblical doctrine of origins of course is foundational to all other doctrines, and if this could be refuted, or even diluted, then eventually the other doctrines of Biblical theology would be undermined and destroyed.

The powerful argument from design in nature, as evidence for God and His creation, would need to be explained by some other means, some naturalistic means, before evolution could really become acceptable to most people. And such a new explanation would need to be a "scientific" explanation, sufficiently so to convince the scientific community that it would really explain evolution. If the scientists could be converted to evolution, then the science-honoring public would soon go along, especially in view of man's basic tendency to rebel against God anyhow.

For a while, it seemed that Lamarck's theory of evolution by the inheritance of acquired characters would serve the purpose. The theory had a superficial appearance of plausibility and did seem to provide an alternate explanation for the evidences of creative design in nature. Lamarck, with his own bitter hatred of the Bible and Christianity, argued his theory very forcefully and persuaded many people of its value. Karl Marx and his colleagues followed Lamarck to some extent and their successors continued to impose it on communist biology until very recent times.

To most people, however, the idea of inheritance of acquired characters was so contrary to all experience that they could never really take it seriously, much as they might like to believe it. Consequently, a better theory was urgently needed, one that would both commend itself to scientists and also be simple for the average man to understand and somehow in keeping with his own common-sense experience.

The idea of natural selection in the struggle for existence was the perfect solution. Everyone was familiar with the effectiveness of artificial selection in breeding, so why wouldn't the same process work in nature? Add the factor of the great spans of geologic time conveniently provided by Lyell's uniformitarianism, and everything was present to explain away the evidence of design and even the real necessity of a Creator. Or at least this was the way it would seem, and that was all that was necessary.

The time was ripe for the Darwinian theory. As noted above, it really wasn't Darwin's theory, but he was the one who was advocating it at the time when it became propitious to renounce Lamarckianism and adopt natural selection instead. Huxley, Lyell, and others prodded Darwin to publish his book, which he had been painfully working on for many years, and it soon became famous. Huxley, along with Spencer, Haeckel, and others immediately opened a relentless evolutionary propaganda campaign, and it wasn't long until essentially the whole world was converted to evolution.

One may sense from the foregoing sequence of events intimations of ominous undercurrents contributing to them. The coincidences seem so improbable, and the results so far-reaching, one can hardly avoid wondering whether the factors culminating in these developments may not have involved more than mere accidents of history.

Before Darwin

It is clear that there were many evolutionists before Darwin. In fact, most of the standard present-day textbook "evidences" for evolution were published and well known long before the *Origin of Species*. However, these views were not predominant, even in the educational world, until the rise of Darwinism. The influence of Christianity had for a long time relegated pagan philosophies to a sort of intellectual underground and the Biblical cosmology was generally accepted in the western world, both by scholars and laymen.

There is no doubt, however, that pagan and gnostic philosophies had exerted some influence on Christian thought, even at the very beginning of church history. These influences are reflected in the frequent uses of allegorical methods of Biblical exegesis, attempting to harmonize Scripture with current cosmological and philosophical notions, and in the frequent lapse of Christian mystics into a sort of pious pantheism. Both tendencies are often accompanied by evolutionary speculations of various kinds.

"During this period it is significant that several of the church fathers expressed ideas of organic evolution even though the trend of ecclesiastical thought led more readily into other lines of reasoning. St. Gregory of Nyssa (331-396 A.D.), St. Basil (331-379 A.D.), St. Augustine (353-430 A.D.), and St. Thomas Aquinas (1225-1274 A.D.) expressed belief in the symbolical nature of the Biblical story of creation and in their comments made statements clearly related to the concept of evolution."[1]

It is true, no doubt, that the dominant point of view in Christendom during these years was in support of literal creation. At the same time, it was not true as often charged that the concept of a stationary earth and geocentric universe, as held by many of these writers, originated in the Bible. The Bible teaches neither of these things, but the church of those centuries was also largely dominated by the philosophy of Aristotle, and these ideas were part of his system, as well as that of Ptolemy.

It is generally recognized that the church of the post-Apostolic period down to the very time of the Reformation, was much influenced by gnosticism, and by other forms of Greek philosophy. The Christian scholars of those centuries, exactly as many at the present time, felt it was essential to work out a compromise cosmology which would be acceptable to the intellectuals of their day. This they did by the simple expedient of interpreting the Scriptures allegorically, which techniques of course can easily convert meanings into anything one desires.

"This desire to find allegories in Scripture was carried to excess by Origen (185-254) who was likewise associated with Alexandrian thought, and he managed thereby to get rid of anything

---

[1] Arthur Ward Lindsey: *Principles of Organic Evolution* (St. Louis, The C. V. Mosby Co., 1962), p. 21.

which could not be harmonized with pagan learning, such as the separation of the waters above the firmament from those below it, mentioned in Genesis, which he takes to mean that we should separate our spirits from the darkness of the abyss, where the Adversary and his angels dwell."[1]

As far as the gnostics themselves were concerned, these were fragmented into many different sects, and it is difficult to generalize about all of them. Most were characterized by an oriental dualism, and believed in a sharp differentiation between the world of the spirit and the world of matter. The latter was believed to be eternal, rather than specially created, and this of course is the main pillar of an evolutionary system. The personal incarnation and bodily resurrection of Christ were denied, as well as many of the other cardinal doctrines of Christianity.

For the most part, as noted, Christians both in the dominant state churches and in the various smaller groups outside these churches retained their faith in the Biblical view of creation. This conviction was especially strengthened by the great increase in Bible circulation following the Reformation and later spiritual awakenings.

As a minority belief, however, evolution was not uncommon. Spontaneous generation had been accepted as common knowledge at least since the time of Aristotle, and was opposed only by a minority of Christians who recognized it as unscriptural. Ideas of transmutation were also widely held, even in the realm of inorganic materials, as evidenced by the studies of the alchemists.

Two levels of evolutionary beliefs need to be recognized. At the intellectual level Greek atomistic philosophies, such as worked out by Democritus and Leucippus, were highly developed and were accepted by many scholars. The pagan mystery religions were understood and practiced by initiates on a considerably higher plane of sophistication than the popular idol worship of the masses. The Stoic and Epicurean philosophers, the best known of which is probably Lucretius, were essentially evolutionary pantheists. None of the pagan religions or philosophies held any real

---

[1] J. L. E. Dreyer: "Medieval Cosmology," *Theories of the Universe,* Ed. by Milton K. Munitz, (Glencoe, Illinois, The Free Press, 1957), p. 117.

belief in a personal, omnipotent, eternal Creator, who created all things out of nothing. The philosophy of Aristotle, who did teach a quasi-creation doctrine while simultaneously advocating spontaneous generation, the philosophy of the Stoics, and the various gnostic philosophies, all had significant influence in keeping the pagan cosmologies alive even in the Christian churches. With the Renaissance came a great revival of pagan philosophies and these came to real fruit in the full-blown evolutionary cosmologies of Kant and LaPlace, with their nebular hypotheses of Descartes, with his mechanistic philosophy, of Spinoza and others.

On the popular level, the philosophical pantheism of the scholars was expressed in the pantheon of gods and goddesses associated with the pagan religions. When these were finally replaced by Christianity, they went underground, as it were. To some extent the idols were "baptized" with Christian names, and the old polytheistic nature-worship incorporated into the customs and practices of the churches. At the same time the demonic and occultic aspects of paganism were perpetuated in various forms of witchcraft, which continued to thrive throughout the Middle Ages and even to modern times.

All of these systems, philosophical pantheism, popular polytheism and occult supernaturalism, are fundamentally evolutionary systems. All were and are bitter opponents of Biblical Christianity, rejecting any concept of a personal God, who created all things, including the physical universe itself, by special creation *ex nihilo*. It is from such as these that the modern theory of evolution must trace its ancestry. Diverse though the "gods many" may be in details, they are all one in their hatred of the true God of creation.

The Pre-Christian World

The most influential of the Greek philosophers was Aristotle (B.C. 384-322). He, of course, was a student of Plato (B.C. 427-348), and he of Socrates (B.C. 470-399). It has been argued whether these men were evolutionists or creationists, and there is no doubt that they did believe in God, a First Cause, or Prime Mover. Their cosmogonies are not too specific in detail.

However, it is well known that Aristotle believed in spontaneous generation, and this is certainly a form of evolution. Furthermore, he believed in the concept that the world never had a beginning, which doctrine of course is inconsistent with true creationism.

"Like his master Plato, Aristotle insists there is but one world, that is a central body like the earth surrounded by a finite number of planets and stars. This one world, of course, which makes up the entire universe, contains all existent matter. . . . Aristotle argues that the one world or universe we know is eternal, without beginning and without end."[1]

Even earlier than these philosophers, however, a much more consistently pantheistic and evolutionary view of the universe was widely believed, and continued to exercise profound influence on all subsequent scientific thought even into modern times. This was the "atomistic" school.

"The type of thinking initiated by the Milesian school of pre-Socratic thinkers — Thales, Anaximander, and Anaximenes — in the sixth century B.C. was carried forward in many directions. One of the most remarkable outcomes of such speculations, representing a culmination of their materialistic thought,was to be found in the Atomist school.Originally worked out in its main features by Leucippus and Democritus in the fifth century B.C., the teachings of atomism were later adopted as a basis for the primarily ethical philosophy of Epicureanism. . . . it elaborates the conception of a universe whose order arises out of a blind interplay of atoms rather than as a product of deliberate design; of a universe boundless in spatial extent, infinite in its duration and containing innumerable worlds in various stages of development or decay. . . . It was the same conception, however, which once more came into the foreground of attention at the dawn of modern thought and has remained up to the present time an inspiration for those modes of scientific thinking that renounce any appeal to teleology in

---

[1] Milton K. Munitz: *Theories of the Universe* (Glencoe, Illinois, The Free Press, 1957), pp. 63, 64. Munitz was Professor of Philosophy at New York University.

the interpretation of physical phenomena."[1]

Modern evolutionary materialists are not so modern after all. Their system is essentially the same as the pre-Socratic Greek cosmology of 2500 years ago! The system persisted through the Roman period, with one of its leading exponents the Roman poet Lucretius. A typical excerpt from his writings follows:

"Certainly the atoms did not post themselves purposefully in due order by an act of intelligence, nor did they stipulate what movements each should perform. As they have been rushing everlastingly throughout all space in their myriads, undergoing myriad changes under the disturbing impact of collisions, they have experienced every variety of movement and conjunction till they have fallen into the particular pattern by which this world of ours is constituted. This world has persisted many a long year, having once been set going in the appropriate motions. From these everything else follows."[2]

The still earlier Greek philosophers, beginning apparently with Thales (640-546 B.C.), were also evolutionists and materialists. Thales' home was Miletus, and his followers are called the Milesian school, and also the Ionian school.

"The Milesian system pushed back to the very beginning of things the operation of processes as familiar and ordinary as a shower of rain. It made the formation of the world no longer a supernatural, but a natural event. Thanks to the Ionians, and to no one else, this has become the universal premise of all modern science."[3]

Their concept of evolutionary development was not, of course, exactly that of modern Darwinism, but the essentials are there. They believed that:

" — the order arose by differentiation out of a simple state of things, at first conceived as a single living substance, later by

---

[1] *Ibid,* p. 6.

[2] Lucretius: *The Nature of the Universe* (Translated by R. E. Latham, New York, Penguin Books, 1951), p. 58. Lucretius lived from 96 to 55 B.C.

[3] F. M. Conford: "Pattern of Ionian Cosmogony" in *Theories of the Universe* (Ed. by Milton K. Munitz, Free Press, 1957), p. 21.

the pluralists, as a primitive confusion in which 'all things,' now separate, 'were together.' "[1]

Thales was influenced in his thinking very much by the Egyptians and the Phoenicians, as well as by the cosmogonic myths of the Greeks themselves, especially as recorded by Hesiod. The scope of this study does not warrant a detailed tracing of the cosmogonies of all the ancient peoples — Persians, Syrians, Egyptians, Canaanites and others. If such were done, however, it could be shown that all of them, in one way or another, were essentially evolutionists, with the one exception, of course, of the Hebrews. As one illustration, however, let us consider another great nation, in a completely different part of the world, also with a long history, namely, China.

"In contrast to the Western world, the Far Eastern philosophers thought of creation in evolutionary terms. . . . The striking feature of the Chinese concept of cosmogony is the fact that creation was never associated with the design or activity of a supernatural being, but rather with the interaction of impersonal forces, the powers of which persist interminably."[2]

Some of the Chinese speculations, in fact, go into amazing detail concerning the supposed sequences of organic evolution, proceeding organism by organism from the simplest plants upward through grubs, insects, birds, leopards, horses, and men.

"Though completely fanciful, this ladder of nature is noteworthy because it was conceived more than two millennia before the Western world began to re-examine its Biblical chronology. But beyond this, the above-quoted passage contains two highly important points: first, a belief in an inherent continuity of all creation and, second, a reference to the merging of one species into another — from primordial germ to man."[3]

The ancient cosmogonies of India also are evolutionary. All of these systems regard the earth as extremely old, perhaps infinite in age. Those modern writers who like to boast of the "discovery" that the earth is billions of years in age instead of thousands as a

---

[1] *Ibid*, p. 22.
[2] Ilza Veith: "Creation and Evolution in the Far East," in *Issues in Evolution*, Ed. by Sol Tax (Chicago, University of Chicago Press, 1960), pp. 1, 2.
[3] *Ibid*, p. 7.

great achievement of modern science are simply ignorant of the almost universal belief of the ancients in evolution and the great antiquity of the earth.

In the light of these facts, the common charge that Moses wrote Genesis in terms of special creation, rather than the more sophisticated concept of evolution, as an accommodation to the naive culture of the Hebrews to whom he was writing, is itself seen to be inexcusably naive! The only cosmogonies his people could have encountered in the world of their day, apart from Genesis, were evolutionary cosmogonies. The concept of a special, recent creation of all things, was a radical, new concept, and would have to be plainly and definitely set forth in the clearest terms, in order for them to grasp it.

Not only do evolutionary systems appear among all the ancient philosophies and religions, however. In spite of many differences in detail, it is well known that the very religions themselves are all essentially the same. Whether in Greece, Rome, Egypt, Canaan, India, or anywhere else, the basic systems seem to be equivalent to each other. Each involves an array of gods and goddesses representing various aspects of nature and life, altogether comprising the great world-spirit which is essentially the personification of the universe itself. The gods and goddesses in each nation always seem to have exact counterparts in the pantheons of other nations, and the ritualistic systems, especially the "mysteries" imparted to their respective initiates, are likewise equivalent.

The classic work of Alexander Hislop, *The Two Babylons,* has never been answered. In this work, Hislop documents from an abundance of sources the primeval unity of the various pagan religions, and traces their origin back to the first Babylon. He says:

"These mysteries were long shrouded in darkness, but now the thick darkness begins to pass away. All who have paid the least attention to the literature of Greece, Egypt, Phoenicia, or Rome are aware of the place which the "Mysteries" occupy in these countries, and that, whatever circumstantial diversities there might be, in all essential respects these "Mysteries" in the different countries were the same. Now, as the language of

Jeremiah (51:7) would indicate that Babylon was the primal source from which all these systems of idolatry flowed, so the deductions of the most learned historians, on more historical grounds, have led to the same conclusion."[1]

In any case, it is evident that the Babylonian cosmogony must be very ancient, and must have had profound influence on those of other ancient nations. In modern times, it has been rediscovered by archaeologists, and is known as *Enuma Elish*.[2] In the form now available, it is believed to have been written about 2600 B.C., and thus to have been written prior to Moses' time. Many people, therefore, have erroneously concluded that Moses borrowed his creation account from this source, but the truth is that the latter represents at best a corrupted form of the true record which was handed down to Moses by the early patriarchs.

Actually, *Enuma Elish* is the Babylonian cosmogonical myth, and like all the others, is an evolutionary system. It was from such as these that the later philosophers such as Thales sought to divest the true materialistic history of the earth.

"One evidence of the influence of myth upon these earliest instances of 'scientific' thought is to be found in the interest in formulating a complete cosmogony which would show how from some primordial state an ordered world arose and underwent successive differentiations of an astronomic, geographic, and meteorologic kind, culminating ultimately, in the mergence of living things and human society."[3]

All such myths began with matter in some form already in existence, and then the forces of nature (or the gods who personify those natural forces) are described as operating upon this primeval matter in such ways as to bring a cosmos out of the chaos. The early philosophers then later tried to modify these original cosmogonies by placing the popular animism and polytheism in a scientific framework of some sort, proposing the

[1] Alexander Hislop: *The Two Babylons* (American Edition, New York, Loizeaux Brothers, 1950), p. 12. This book was originally published in Edinburgh in 1858, and has gone through numerous editions and printings.

[2] Alexander Heidel: *The Babylonian Genesis* (University of Chicago Press, 1951).

[3] Milton K. Munitz: *Space, Time, and Creation* (Glencoe, Illinois, The Free Press, 1957), pp. 8, 9.

development of things from primeval fire or water or atoms or some other primordial stuff.

Specifically, *Enuma Elish* assumes that all things have evolved out of water.

"This description presents the earliest stage of the universe as one of watery chaos. The chaos consisted of three intermingled elements: Apsu, who represents the sweet waters; Ti'amat, who represents the sea; and Mumsu, who cannot as yet be identified with certainty but may represent cloud banks and mist. These three types of water were mingled in a large undefined mass. . . . Then, in the midst of this watery chaos, two gods came into existence — Lahau and Lehemu." [1]

Then the epic goes on to describe how other gods were generated and then engaged in various activities, including fighting and killing one another. Eventually the god Marduk gains control of the heavenly host and thence proceeds to the formation of the earth and stars and man, the latter actually from the blood of one of the slain gods.

The complicated battles of the gods and goddesses seem clearly to portray the struggling forces of nature as they labor to bring forth an orderly world. Or, perhaps, they may rather represent actual warfare in the heavens, between Satan and his angels and Michael and his angels. Or possibly both.

In any case, such tales are certainly infinitely inferior to the true record of creation as given in the first chapter of Genesis. The idea that Genesis could have been derived from such as this is incredible.

We have thus shown that the evolutionary philosophy is not modern at all, but rather traces back through all the history of mankind, right back to Babylon — not the Babylon of Nebuchadnezzar (though it was prominent there) but to the original Babel founded by Nimrod (Genesis 10:8-10). Furthermore, the evolutionary system is part and parcel of the system of pantheistic polytheism which constituted the universal religion of the ancients, and which also was derived from Babel.

Furthermore, this system is invariably identified in some way

---

[1] Thorkild Jacobsen: "Enuma Elish — the Babylonian Genesis," in *Theories of the Universe* (Edited by Milton K. Munitz, The Free Press, 1957), p. 9.

with astrology, and all the various divinities are associated with their own particular stars or planets. To the pagans, these heavenly beings were not considered as mere religious ideals, but as living spirits, capable of communicating directly with men through oracles or seers or mediums. Pantheism, polytheism, astrology, idolatry, mysteries, spiritism, materialism — all this great complex of belief and practice, superficially diverse, but fundamentally one — constitutes the gigantic rebellion of mankind against the true God of creation. Always, whatever the outward appearance, the underlying faith is in eternal matter, in a self-contained cosmos evolving upward out of chaos toward future perfection. Though in some cases, particularly isolated tribes, the people seem to have retained some kind of dim awareness of a great High God, far removed in time and space from their personal lives, their practical interests have from primeval times invariably been centered in the various divinities connected with their own immediate environments. Such identification of ultimate reality with finite natural objects is nothing but evolution. Matter in some form, not God, is their original and eternal cause of all things.

The Beginning of Evolution

The origin of evolution as a religious philosophy (and that, of course, is all that it can ever be) is thus locked together with the origin of paganism, which in the post-diluvian world was undoubtedly at Babel. This is also intimated by Scripture, when it speaks of "Mystery, Babylon the Great, the mother of harlots and abominations of the earth" (Revelation 17:5). Since Nimrod was the founder and first ruler of Babylon, it seems reasonable to propose that he was responsible for the introduction of this entire system into the life of mankind.

To the modern skeptic, of course, Nimrod is merely a nonexistent legendary hero like all the other names recorded in the early chapters of Genesis. Such skeptics should at least realize, however, that the literature of antiquity frequently refers to Nimrod in one way or another. There were many people in ancient times, in addition to the writer of Genesis, who regarded him as real.

There are numerous ancient historians who recognize Nimrod and his exploits, and various sites in the Babylonian-Nineveh region are associated by the Arabs with his name. It is quite probable that the chief God of the Babylonians (Marduk, or Merodach), really represents the same Nimrod, deified after his death.

Obviously, however, when we attempt to decipher history of this degree of antiquity, we have little to go on. This is the period in which fact and legend are almost indistinguishable. Although a great deal of archaeological excavation has been done in the Tigris-Euphrates valleys, the monuments are difficult to translate, and only a small fraction even of the recovered materials has really been read. Much still remains to be excavated, and of course the far greater part has long since been destroyed by the ravages of time.

We do, of course, have the record of Genesis 10 and 11. Though these chapters are tantalizingly brief, we can have confidence that the information they give is true. To the skeptic, we can at least say two things — no one has disproved the validity of these verses, and he has as yet nothing better to offer.

In this section, as we seek to understand the ultimate source of the evolutionary system, speculation is admittedly necessary. But at least it is speculation guided by the information given in Scripture, as well as by the historical data we have been accumulating.

We assume, therefore, that the Babylonian mysteries were originally established by Nimrod and his followers at Babel. They have somehow since been transmitted throughout the world and down through the centuries, corrupting all nations with their materialistic glorification of the "host of heaven," changing the "glory of the uncorruptible God into an image like to corruptible men, and to birds, and fourfooted beasts, and creeping things" (Romans 1:23). Because they "did not like to retain God in their knowledge" (Romans 1:28), they proceeded to change "the truth of God into a lie, and worshipped and served the creature more than the Creator" (Romans 1:25).

The remarkable similarities and antiquities of the Zodiacal

constellations and the astrological systems that have come down from all the early nations, provide strong evidence of the primeval unity of heathendom. It therefore is a reasonable deduction, even though hardly capable of proof, that the entire monstrous complex was revealed to Nimrod at Babel by demonic influences, perhaps by Satan himself.

It is significant that the phrase "the host of heaven" is applied in Scripture both to the stars and to angels. Similarly, the worship of the sun, moon, and stars, as well as the mythological deities, and the graven images which represent them, is also frequently identified in Scripture with the worship of angels, especially with the fallen angels and the demonic hosts who are following Lucifer, the "day-star" (Isaiah 14:12-14; Revelation 12:4) in his attempt to replace God as king of the universe.

In common with all the other great temple-towers of antiquity, it is likely that the original Tower of Babel (Genesis 11:4) was built, not to "reach unto heaven" in a literal sense (Nimrod was no naive character in a fairy tale to attempt such a thing as that) but rather with a "top unto heaven" (the words "may reach" are not in the original). That is, its top was a great temple shrine, emblazoned with the zodiacal signs representing the host of heaven, Satan and his "principalities and powers, rulers of the darkness of this world" (Ephesians 6:12). These evil spirits there perhaps met with Nimrod and his priests, to plan their long-range strategy against God and His redemptive purposes for the post-diluvian world. This included especially the development of a non-theistic cosmology, one which could explain the origin and meaning of the universe and man without acknowledging the true God of creation. Denial of God's power and sovereignty in creation is of course foundational in the rejection of His authority in every other sphere.

The solid evidence for the above sequence of events is admittedly tenuous. As a hypothesis, however, it does harmonize with the Biblical record and with the known facts of the history of religions; whereas, it is difficult to suggest any other hypothesis which does.

If something like this really happened, early in post-diluvian history, then Satan himself is the originator of the concept of evolution. In fact, the Bible does say that he is the one "which deceiveth the whole world" (Revelation 12:9) and that he "hath blinded the minds of them which believe not" (II Corinthians 4:4). Such statements as these must apply especially to the evolutionary cosmology, which indeed is the world-view with which the whole world has been deceived.

One question remains. Assuming Satan to be the real source of the evolutionary concept, how did it originate in his mind? He originally was "full of wisdom and perfect in beauty . . . in the day that thou wast created" (Ezekiel 28:12, 13). Surely, he knew full well that he, as well as all other angels and everything else, had been created by God, not evolved by natural processes. Knowing this, he could never have the slightest hope of succeeding in the cosmic rebellion he has been promoting for so many millennia.

A possible answer to this mystery could be that Satan, the father of lies, has not only deceived the whole world and the angelic hosts who followd him — he has even deceived himself! The only way he could really *know* about creation (just as the only way *we* can know about creation) was for God to tell him! But "thine heart was lifted up because of thy beauty, thou hast corrupted thy wisdom by reason of thy brightness" (Ezekiel 28:17).

The sin of pride and unbelief, the twin source of all other sins, resulted in "the condemnation of the devil" (I Timothy 3:6). He refused to believe and accept the Word of God concerning his own creation and place in God's economy. Perhaps resentment at the creation of man in God's image, with the marvelous ability to multiply his own kind (which angels cannot do) and the commission as God's vice-regent over the earth (infinitely more beautiful and complex than all the stars of heaven) contributed to the development of this unbelieving pride in Satan. He therefore deceived himself into supposing that all things, including himself and including God, had been evolved by natural processes out of the primordial stuff of the universe, and that therefore he himself might hope to become God.

He therefore said, ''I will be like the most High'' (Isaiah 14:14) and God ''cast him to the earth'' (Ezekiel 28:17). He then brought about man's fall with the same deception (''ye shall be as gods'') and the long sad history of the outworking of human unbelief as centered in the grand delusion of evolution has been the result.

Perhaps the ''Babylonian Genesis,'' the *Enuma Elish,* represents not only the deceptive evolutionary cosmology which he taught Nimrod, but also the primeval Lie with which he deceived himself. According to the Bible, the initial state of the created universe was one of a pervasive watery matrix, in which the elements of the unformed ''earth'' were everywhere in suspension or solution. It was *this* universe which Satan saw as he first came into his divinely created existence on the first day of the creation week (Psalm 104:4 indicates that God created the angels directly after He had created the heavens and light and His own chambers in the heavens).

When Satan later wished to find another explanation than God for the universe, he would have to hark back to the primordial state of the universe as he had first encountered it, and then assume this watery chaos (note the description in the *Enuma Elish* as quoted on page 71) to be the ultimate reality out of which both he and God had somehow evolved. Thus he himself became the patriarchal Evolutionist, and the ultimate originator of all later systems of evolution, some very naive and fanciful, some very sophisticated and apparently scientific.

No doubt the above suggested conclusion regarding the Satanic origin of evolution will be angrily derided by all atheistic, humanistic, or pantheistic evolutionists. They believe in neither God nor Satan, and so this inference will seem absurd to them.

For those who *do* believe in God and Satan, however, (as did the Lord Jesus Christ), one is almost necessarily carried to such a conclusion by the combined witness of Scripture, history, and logical deduction.

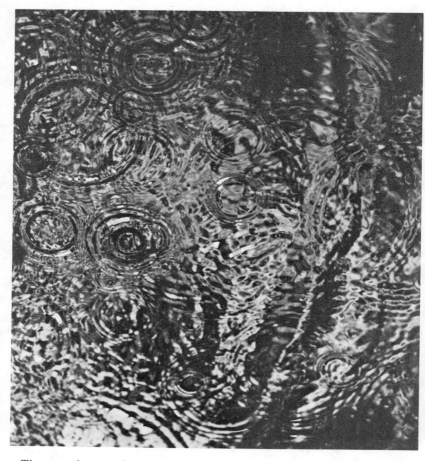

The popular myth that evolution is the scientific explanation of how things began, while creation is merely a religious belief, needs correction. The "evidence" for evolution is really only the assumption of evolution, and scientists who support evolution are merely reasoning in circles. The real scientific evidence supports creation!

Raindrops on Pond

Photo credit: Tim Ravenna

CHAPTER IV

IN SCIENTIFIC CIRCLES

How to Define Evolution

Dialogue between evolutionists and creationists has long been stymied by emotional semantics. Whether this terminological confusion has been intentional or accidental, time will judge. Now, however, it is time to define terms carefully and to avoid imputation of ulterior motives on either side, at least as far as individual personalities are concerned. Creationists need not arbitrarily categorize evolutionists as atheists (many evolutionists certainly believe in some kind of God — even though evolution as a system is fundamentally anti-theistic). Nor should evolutionists dismiss creationists as ignoramuses (although many creationists are admittedly indifferent or antagonistic toward science, there are today thousands of qualified scientists and other well-educated people who have rejected evolution for scientific as well as religious reasons).

It would help, for example, if evolutionists would quit citing the small changes observable in the present as proof of large changes in the past. Creationists are well aware of the former, but balk at the leap of faith required for the latter.

Both groups should recognize that the subject of origins is ultimately beyond the scope of empirical science. The essence of the scientific method is experimental repetition, and one cannot repeat the origin of the solar system or the origin of man in his laboratory.

There is no experiment that can be devised which can discriminate between total evolution and creation. These, therefore, are not matters of science at all. They cannot even be compared by canons of historical investigation, since they took place before the advent of written records.

The outstanding philosopher of science, Karl Popper, though himself an evolutionist, pointed out cogently that evolution, no less than creation, is untestable and thus unprovable.

"Agreement between theory and observation should count for nothing unless the theory is a testable theory, and unless the agreement is the result of attempts to test the theory. But testing a theory means trying to find its weak spots; it means trying to refute it. And a theory is testable if it is refutable. . . .

"There is a difficulty with Darwinism. . . . It is far from clear what we should consider a possible refutation of the theory of natural selection. If, more especially, we accept that statistical definition of fitness which defines fitness by actual survival, then the survival of the fittest becomes tautological and irrefutable."[1]

Neither evolution nor creation can be considered a scientific fact since neither can be proved scientifically. Neither evolution nor creation can be regarded as a legitimate scientific theory, since neither can be tested scientifically to see whether or not it is true. The small changes that can be observed (e.g., mutations in *Drosophila,* island-to-island variations in finches, changes in coloration in the peppered moth, etc.) fit well in *either* "theory," and so are irrelevant as discriminants.

---

[1] Karl R. Popper: "Science: Problems, Aims, Responsibilities," *Proceedings, Federation of American Societies for Experimental Biology,* Vol. 22, 1963, p. 964.

Is there no way, then, that the two views can be evaluated and compared scientifically? Yes, the proper and legitimate, non-emotional[1] approach is to set up two carefully-defined scientific models — an evolution model and a creation model — and then to use the two models on a comparative basis for the prediction of observable data. Though neither is "falsifiable" and neither is "demonstrable," since they ultimately relate to pre-history, their effectiveness in predicting data can be objectively compared. The decision between the two may still be somewhat subjective, but at least it will be informed by a scientific comparison of facts.

Evolutionists believe in general that the evolutionary process is universal in scope, accounting for the origin and development of elements, planets, galaxies, proteins, cells, plants, animals, man and, indeed, all things. Huxley defined evolution in the following fashion:

"Evolution in the extended sense can be defined as a directional and essentially irreversible process, occurring in time, which in its course gives rise to an increase of variety and an increasingly high level of organization in its products. Our present knowledge indeed forces us to the view that the whole of reality is evolution . . . a single process of self-transformation." [2]

With a similarly broad brush, Dobzhansky treats evolution as follows:

"Evolution comprises all the stages of the development of the universe: the cosmic, biological and human or cultural developments. Attempts to restrict the concept of evolution to biology are gratuitous. Life is a product of the evolution of inorganic nature, and man is a product of the evolution of life."[3]

Rene Dubos says:

"Most enlightened persons now accept as a fact that everything in the cosmos — from heavenly bodies to human beings — has

---

[1] Why is it that whenever scientific questions are raised concerning the validity of evolution, evolutionists respond angrily, or sarcastically, rather than scientifically? Is there a "religious" commitment to evolution?

[2] Julian Huxley: "Evolution and Genetics," Chapter 8 in *What is Man?*, Ed. by J. R. Newman (New York, Simon and Schuster, 1955), p. 272.

[3] Theodosius Dobzhansky: "Changing Man," *Science,* Vol. 155, No. 3761, January 27, 1967, p. 40.

developed and continues to develop through evolutionary processes."[1]

Since all will recognize that Huxley, Dubos, and Dobzhansky are three of the world's leading evolutionists, we can safely define the evolution model in some such terms as they have outlined. Undoubtedly, most other evolutionists would agree with this general perspective.

Contrast with the Creation Model

The creation model does not need to be formulated in "religious" terminology, evolutionists to the contrary, notwithstanding. It states simply that the major categories of nature were formed by special creative and cataclysmic, purposive processes in the past which are no longer in operation today and which therefore are not accessible to empirical observation.

The evolution model, on the other hand, postulates that all things can be explained in terms of processes which operated in the past and are still continuing in the present. Thus, by extrapolation of present processes, it should be possible to describe the origin and development of all natural categories and phenomena.

Note that in neither model is there any necessary reference to religious concepts. Neither model refers to the Bible nor to any specific dates in the past. There is no reason, therefore, to reject evolution as "anti-Biblical" or to reject creation as "pro-Biblical." As the models have been formulated, neither necessarily has any relation to the Bible at all, pro or con.

The evolution model merely states that all things *can* be explained in terms of present processes; the creation model states that all things *cannot* be explained in terms of present processes. More specifically, the creation model stipulates one or more periods of *special creation* and one or more periods of *special destruction* in the past, when processes were operating which are fundamentally incommensurate with those which are functional today. No religious or theological implications need be suggested for either of these, insofar as scientific comparison of the two models is concerned.

---

[1] Rene Dubos: "Humanistic Biology," *American Scientist,* Vol. 53, March 1965, p. 6.

Surely both evolutionists and creationists can forego, for the time being, the tendency either to recoil from the theological implications or to rejoice in the theological implications, of the respective models, as the case might be, until the two models have been first compared objectively and scientifically. Once one or the other has been shown to be superior, then the theological inferences, if any, can be drawn.

Admittedly such objectivity is hard to achieve when dealing with models of origins. For some reason there are such emotional overtones in this problem that anger or sarcasm tend to surface very easily, on both sides, and this must be avoided if any progress in understanding is to be attained.

What we propose is to take the two models in their simplest forms, as outlined above, and to use them on a comparative basis to predict and correlate various types of observable data. Their respective predictions can then be tested against observations.

Similarities, Differences, and Taxonomic Classification

The obvious place to begin in this comparison of model predictions is in the array of living plants and animals in the organic world. This array is the present product of any past processes of evolution or creation, and so certainly should provide some clue to the nature of these processes.

However, to be truly objective at this point, we need to try as well as we can to purge our minds of any prior knowledge we possess as to the characteristics of these organisms. We must try to make predictions strictly on the basis of the two models, and then, as it were, venture forth for the first time into the real world to find what is actually there.

The evolution model attempts to explain the entire organic assemblage in terms of natural descent from a common ancestor. Therefore the most obvious prediction from this model would seem to be that all such descendants should be alike. Having come from the same ancestor by the same processes in the same world, there is no immediately apparent reason why any one of these descendants should be different from any other.

It is not quite so simple, however. As the descendants multiply, they must necessarily occupy more space and, eventually, this space may become large enough to encompass more than one type of environment. Assuming now that the hypothetical common ancestor somehow had the genetic capacity to vary from one individual to another (the origin of this variegated genetic "information" also needs to be explained, of course), then the gradually changing environments should elicit gradually diversifying descendants responding to those environments.

Thus the organic assemblage predicted from the evolution model need not be one of uniform sameness. However, it does seem that the model predicts at least a *continuum* of all forms of life. There exists a continuum of environments and a common ancestor and a common process of development. Therefore there should be a continuum of organisms, and a classification system would be impossible.

That this is necessarily the fundamental prediction of the evolution model is obvious from the fact that the discrete gaps between the various kinds of organisms require further explanation. If there were no gaps, but only a continuum, no other explanation would be needed. Such a continuum, if it existed, would be properly considered an exceptionally clear evidence of past evolution.

The creation model, on the other hand, in its basic form, predicts that special creation, being purposive, would result in a discrete array of clear-cut distinct organisms, each with its own peculiar structure provided for its own particular functions. There would be many similarities, but also many differences. The creative process would have designed similar structures for similar functions and different structures for different functions. Since both fish and men would have need to see, eyes would be provided for both, but fish would receive gills and men would receive lungs, corresponding to the particular environments in which they were created to function.

What, then, do we actually find in nature, a continuum of organisms, or an array of clear-cut kinds? Let the noted evolutionary geneticist Dobzhansky give the answer.

"If we assemble as many individuals living at a given time as we can, we notice at once that the observed variation does not form any kind of continuous distribution. Instead, a multitude of separate discrete distributions are found. In other words, the living world is not a single array of individuals in which any two variants are connected by unbroken series of intergrades, but an array of more or less distinctly separate arrays, intermediates between which are absent or rare."[1]

In the evolution model, similarities (whether in anatomy, embryology, blood chemistry, or whatever) are predicted on the basis of common ancestry. In the creation model, the same similarities are predicted on the basis of a common purposive designer. Gaps and differences are likewise predicted by the creation model as the product of purposive design. Gaps and differences are not predicted at all by the evolution model, except on the basis of subsidiary hypotheses that must be introduced for this specific purpose. Thus the very existence of a science of taxonomy is a prediction of creationism and a problem to evolutionism.

The Unprogressive Nature of Biologic Change

Having considered the array of organisms, let us next consider the nature of the specific processes which have produced these organisms. Here we need to be more specific than merely to label them evolutionary processes and creative processes.

The evolution model predicts there is some sort of biologic process which impels simple organisms to advance into complex organisms. Particles have become molecules and molecules have advanced to cells and simple cells have progressed to become people. Though the process need not necessarily be continuous, since it has somehow evolved particles into people, it must be there. As our hypothetical innocent scientist, entering for the first time into the world, begins to study its biologic mechanisms, his evolution model would make him expect to observe in action some powerful and pervasive process whose pressures lead inexorably to great advances in complexity and order.

There is nothing in the basic model to tell him whether this process operates slowly or rapidly. The actual tempo of evolution is unknown at first, but it must at least be rapid enough to be

---

[1] Theodosius Dobzhansky: *Genetics and the Origin of Species* (New York, Columbia University Press, 1951), pp. 3, 4.

observable — else, how could it produce such near-infinite results in finite time? Furthermore, if it is not observable, it is beyond the reach of the scientific method. Absence of such a process must be then considered *prima facie* evidence that evolution is not scientific. If it operated exclusively in the past and cannot be observed in the present, then it becomes essentially synonymous with creation.

We must realize, too, that this process must not be one which merely shuffles things around at the same level, or one which may even lower the level of order and complexity. To account for the supposed development of all things from a common ancestor — say a protozoan or simpler — it must be essentially a process which insistently *increases* order and complexity. If we do find such a process clearly and regularly operating in the real world of experimental observation, it can rightly be considered as strong evidence of evolution.

The creation model predicts, on the other hand, that no such process will be observed at all. Since it presupposes that creation of all the basic kinds — including man, the most complex of all — was a completed event of the past, it says explicitly that no natural process of evolutionary development from a simpler kind of organism to a more complex kind of organism can be observed operating today.

It does not predict, however, that there would be no changes at all. On the contrary, it predicts there must be some sort of conservative process which will act to perpetuate the kind in spite of environmental changes. That is, if the environment changes, some process (call it natural selection, if you will) takes over to implement a corresponding change in the species, enabling it to survive as a species in the new environment. The classic example of the peppered moth in England perfectly illustrates this prediction of the creation model. Such modifications are not to be construed as increases of order, but as recombinations, at the same level of order, of the factors built in to the created kind in the beginning.

The creation model would further predict that, since each created kind had its own structure and purpose as it was created, this conservative process would not only tend to maintain its integrity against *external* environmental changes, but also against deterioration due to *internal* causes. That is, if some genetic damage occurred in the kind for some reason (call it a genetic mutation, if you will), then this conservative process (let's go ahead and call it natural selection) would tend to prevent the build-up of the "genetic load" to a level that would endanger the continued existence of the kind itself. That is, the genetically-damaged members of the population would be eliminated, hopefully, before this damage could permeate the entire population. Thus the kind would survive even if individuals and varieties within the kind may die out.

All of the above are clear-cut and direct predictions of the creation model. The model does not, in its basic form, predict mutations, but it does predict variations and also that, if mutations occur, natural selection will tend to eliminate them and thus preserve the integrity of the kinds.

The universally-observed fact that, throughout the period of human observation, "like begets like" is of course consistent with the creation model. The shift from predominant light coloration to predominant dark coloration in the peppered moth, with the increasingly smoky atmosphere and darkening of the tree trunks during the industrial revolution, likewise is a striking confirmation of the creation model. The fact that the thousands of artificially-induced mutations in *Drosophila,* the fruit-fly, have not caused it either to become extinct or to develop into something other than *Drosophila,* likewise conforms to the creation model. To all appearances, there does seem to be a basic conservational principle in nature, as predicted.

As far as the prediction of the evolution model that basic vertical changes in organisms take place in the direction of increasing complexity is concerned, it is well known that mutations are always, or at least almost always, harmful.

"Any chance alteration in the composition and properties of a highly complex operating system is not likely to improve its manner of operation and most mutations are disadvantageous for this reason."[1]

"The truth is that there is no clear evidence of the existence of such helpful mutations. In natural populations endless millions of small and great genic differences exist, but there is no evidence that they arose by mutation."[2]

The apparent absence of a measurably progressive process in nature does not necessarily contradict the prediction of the evolution model that there is such a process, however. It merely requires that an additional assumption be introduced to the effect that the process goes so slowly as to be non-observable within the time-span available for human observations. The variational processes which seem superficially to reflect the basic stability of kinds, and the mutational processes which seem superficially to result in deterioration, are thus assumed in the long term really to represent imperceptibly slow increases in order into new and higher kinds.

The Altogether Missing Links

It is obvious that the fossil record is the most important test of the two models, especially of evolution. Since vast spans of geologic time have been found necessary to allow evolutionary changes of significance, the question of what actually did happen in ancient eras is of primary importance.

The creation model of course must predict that the array of organisms preserved as fossils will correspond to the same classification system as applicable to present-day plants and animals. Since all the basic categories were created in the beginning, these must have persisted ever since and into the present. Therefore, essentially the same kingdoms, phyla, classes, orders, etc., would apply to fossil classification as apply to modern classification. The same sorts of "gaps" between different kinds of organisms in the present world would be anticipated in the fossils.

---

1 Frederick S. Hulse: *The Human Species* (New York, Random House, 1963), p. 53.

2 C. P. Martin: "A Non-Geneticist Looks at Evolution," *American Scientist,* January 1953, p. 101.

The creation model does not preclude extinction of kinds, any more than it precludes the death of individuals, but it does preclude the development of new kinds from older kinds. Therefore, it predicts there will be no true transitional forms, from one kind into another kind, found in the fossil record. There may well be many variations within kinds, including transitions from one variety to another variety, but no true transitional intermediates from one kind to another. It predicts that, whenever a new kind first appears in the fossils, it will already be a fully typical representative of that kind.

These are rather rigid constraints that we have placed on the creationist predictions, and ought to give a clear indication as to its probable validity. No preliminary forms, no transitional forms, clear gaps between kinds, same taxonomic categories as at present, etc., — all of which are explicitly different from the predictions of evolution.

That is, if evolution is true and if all organisms really have gradually developed from a single, primeval, simple ancestral form during the geologic ages, then the evidence for that development ought to be found in the fossil record if it is found anywhere. There must have been many preliminary forms and there must have been many transitional forms and there should at least be a statistical sampling of these preserved in the fossils. Though some gaps certainly are to be expected, because of the accidental nature both of fossilization and of fossil discoveries, such gaps should at least be randomly distributed between the present kinds and the transitional kinds. Gaps in the fossil record should not be the same kinds of gaps as those in the present taxonomic system, and there should be many true transitional intermediates found as fossils — if the evolution model has any value as a predictive system.

In fact, if evolution is true, even the basic taxonomic categories should have been evolving, and there should be evidence of these changes in the fossils.

That these predictions are the obvious and basic predictions of the evolution model can be appreciated if one stops to imagine what the reaction would be if these were actually found. That is, if the fossil record really did yield an abundant sampling of true incipient and transitional forms, if actually there were changing taxonomies, if the fossil gaps were not regular and systematic but only statistical, all of this would be hailed as striking proof of evolution. Such an array of evidence would indeed be hard for the creation model to overcome.

Since, however, this type of evidence is altogether lacking in the fossil record, auxiliary conditions have to be imposed on the evolutionist predictions. It may be assumed, for instance, that evolution took place explosively, in small populations, giving little opportunity for fossilization of the transitional forms. Additional assumptions must be introduced then to account for such evolutionary spurts, since the process today is apparently too slow to observe at all. The sudden appearance of all the taxonomic categories, most of them a half-billion years ago in the Cambrian, with no fossil record of their prior development, is not easy to incorporate into the evolution model, though certain possibilities can be suggested.

No such auxiliary assumptions and conditions are needed for the creation model. All of its predictions are explicitly confirmed by the fossil record exactly as it stands. Even the order of appearance of the various fossils in the geologic strata is anticipated in the creationist framework, but this will be considered later.

It may be well at this point, however, to document (from evolutionary writers) the fact that there are no incipient or transitional forms in the fossils, since it is commonly believed that there are. Paul Moody, in his standard textbook, says:

"So far as we can judge from the geologic record, large changes seem usually to have arisen suddenly, . . . fossil forms, intermediate between large subdivisions of classification, such as orders and classes, are seldom found."[1]

That the dearth of intermediate forms extends even to smaller divisions is confirmed by Davis:

"The sudden emergence of major adaptive types as seen in the abrupt appearance in the fossil record of families and orders,

---

1 Paul A. Moody: *Introduction to Evolution* (New York, Harper & Row, 1962), p. 503.

continued to give trouble. A few paleontologists even today cling to the idea that these gaps will be closed by further collecting, but most regard the observed discontinuity as real and have sought an explanation."[1]

In his day, Darwin attributed these gaps to the limited number of fossils collected. This explanation is no longer adequate.

"There is no need to apologize any longer for the poverty of the fossil record. In some ways it has become almost unmanageably rich and discovery is outpacing integration. . . . The fossil record nevertheless continues to be composed mainly of gaps."[2]

Finally, no less an authority than Simpson acknowledges the regular systematic existence of these gaps in the fossil record:

"In spite of these examples, it remains true, as every paleontologist knows, that *most* new species, genera, and families, and that nearly all categories above the level of families, appear in the record suddenly and are not led up to by known, gradual, completely continuous transitional sequences."[3]

Note that the same classification categories existed throughout the long ages of the fossil record as originally worked out for existing plants and animals. The system itself showed no signs of evolving. Furthermore, most of the major categories (down through at least the families) have remained clearly distinct since their first appearance in the record.

Now perhaps it is true that the gaps are still due to the rarity of fossil deposition and discovery. Or perhaps, as most think, the gaps are due to spurts of explosive evolution caused by periods of intensified cosmic radiation. The fact is, the gaps are still *there,* and this is a primary prediction of the creation model.

Never are fossils of creatures found with incipient eyes, with half-way wings, with half-scales turning into feathers, with

---

1 D. Dwight Davis: "Comparative Anatomy and the Evolution of Vertebrates" in *Genetics, Paleontology and Evolution,* Ed. by Jepsen, Mayr & Simpson (Princeton, N. J., Princeton University Press, 1949), p. 74.

2 T. N. George: "Fossils in Evolutionary Perspective," *Science Progress,* Vol. 48, January 1960, pp. 1, 3.

3 George Gaylord Simpson: *The Major Features of Evolution* (New York, Columbia University Press, 1953), p. 360.

partially-evolved forelimbs, or with any other nascent or transitional characters. Yet there must have been innumerable individuals which possessed such features, if the neo-Darwinian model of evolutionary history is correct. It seems passing strange that the fossilization process selected only those individuals for preservation which already had completed particular stages of evolutionary progress, and yet preserved these in great abundance and variety. Was Nature somehow ashamed of her evolutionary embryos?

The creation model predicts directly that the fossil record will be composed of the same classification categories as those of the present world, with the same kinds of gaps between the categories, and with no evidence of gradual transitions between these categories. This is exactly what is found. Creationists therefore believe that the fossil record, while not impossible of reconciliation with the evolution model, does conform more simply and directly to the creation model.

The Order of the Fossils

There is another aspect of the fossil record, however, which seems to support the evolution model. Different forms of life seem to have first appeared in different geologic ages — first invertebrates, then marine vertebrates, then amphibians, then reptiles, then birds and mammals, then man. Some such sequence as this is of course a primary prediction of the evolution model. Creationism on the other hand would expect to find all the major kinds of organisms appearing at essentially the same time, unless there were a number of different periods of creation.

This latter idea, called by its advocates "progressive creation" is inconsistent with the postulate of a purposive, knowledgeable Creator, who knew what He was doing at the beginning. It usually yields rather readily to the more consistent concept known as "theistic evolution," which supposes that the Creator used evolution as His method of creation and thus energized the whole process when it first began. However, theistic evolution is no different from atheistic evolution, insofar as any scientific criteria are concerned; there is no scientific test that can be proposed to distinguish one from the other. One must choose between theistic and atheistic evolution solely on the basis of theological criteria, not scientific.

Does, then, the sporadic appearance of different forms of life constitute a decisive discriminant against the creation model? Not necessarily. Creationists maintain that, rightly interpreted, the fossil record *does* show that all the forms of life in the fossils existed contemporaneously from the beginning, exactly as predicted by their model.

However, such a reinterpretation requires a critical analysis of the physical character of the sedimentary rocks constituting the matrix of the fossil record, as well as of the fossil deposits themselves. The evolution model of course predicts that these rocks, containing as they do the fossil record of the supposed gradual evolution of life throughout the geological ages, must have themselves been formed slowly and gradually throughout those ages. The doctrine of uniformitarianism has accordingly been applied to their interpretation, attempting to explain them in terms of hydraulic and sedimentary processes as they operate in the present.

The creation model, on the other hand, predicts that since all major kinds of organisms were created concurrently in the beginning, the rocks containing their fossils must have been deposited mostly in one single great epoch of deposition. This prediction in turn requires catastrophism, with the rocks having been originally formed as sediments in a universal hydraulic cataclysm.

Thus this particular comparative test of the creation and evolution models devolves upon whether catastrophism or uniformitarianism more effectively accounts for the sedimentary rocks in the earth's crust, at least those containing fossils.

It would seem that, if the rocks were all formed in essentially the same cataclysmic period, rather than over a long series of distinct ages, they should all exhibit essentially the same physical characteristics. This prediction of catastrophism is precisely fulfilled. Shales, sandstones, granites, and limestones are found indiscriminately in rocks of all the so-called "ages." Lithology, petrology, mineralogy, structure, seismicity, and all other physical characteristics of the rocks apply indiscriminately throughout the geologic column, with the possible exception of the late Tertiary and Quaternary systems, which are believed by most creationists to have been formed after the global catastrophe.

On the basis of the evolution model, requiring a billion years of time for the formation of the fossils, one might expect at least some kinds of physical changes in the rocks over the ages. There seem to be none, however. So far as physical characteristics are concerned, they might all well have been formed in a single epoch.

The existence of fossils also argues for rapid formation of each rock formation containing them. Marine micro-fossils may accumulate slowly on the sea-bottom, but larger fossils normally require rapid burial and lithification in order to be preserved against the inexorable forces of decay. Thus each significant fossil deposit — and to a considerable extent, this applies to all sedimentary rocks — is evidence of at least a local catastrophe of some kind.

The igneous rocks found in the geologic column, both intrusives and extrusives, were of course each formed rapidly since they represent flowing and cooling magmas. Massive limestones and other non-fossiliferous marine complexes are best understood in terms of rapid precipitation. Salt domes and similar massive formations of so-called "evaporites" may best be explained as precipitates, modified by later tectonics, all occurring rather rapidly.

What about the sedimentary rocks themselves? These were once loose sediments, transported and deposited by water. Are they better understood in terms of uniformitarian or catastrophic hydraulic processes?

Consider the following syllogistic chain of reasoning. Each distinct sedimentary stratum in a formation represents a uniform set of hydraulic properties (velocity, flow direction, sediment load, etc.). The next stratum above represents a change in one or more of these properties, thus forming a distinct interface between the two strata. Each of these strata must have been formed quite rapidly, since a constant set of hydraulic properties in a given flow will persist only for a brief time. On the surface of a sedimentary deposit like this will inevitably be formed ripple marks and other irregularities. Such irregularities are commonly noted at the interface between two successive sedimentary strata, thus indicating that the higher stratum was formed very soon after the lower, before such irregularities could be eroded away. The entire series of strata constituting a given formation must therefore have been formed rapidly and essentially continuously.

The top of the formation may well represent an unconformity, with an unknown interval of time, perhaps even involving uplift and erosion or even tilting and truncation, before the next formation is deposited on top of it.

However, there are no worldwide unconformities. Somewhere, if one traces the formation laterally far enough, he will find it grades, either laterally or vertically or both, into another formation, with no unconformity between. Thus the bottom stratum of this second formation must have been laid down immediately after the top stratum of the first.

One then analyzes the strata of the second formation in the same way he did the first. Each stratum was formed rapidly and immediately after the one below it, so that again the entire formation accumulated rapidly. Thus, each formation itself was formed rapidly and immediately following at least one other formation below it.

This process can be traced from bottom to top of the entire geologic column, even through the end of each major geologic system on to the immediate beginning of the next geologic system. There is no time-break between the end of one "age" and the beginning of the next.

"The employment of unconformities as time-stratigraphic boundaries should be abandoned. . . . Because of the failure of unconformities as time indices, time-stratigraphic boundaries of Paleozoic and later age must be defined by time, — hence by faunas."[1]

1 H. E. Wheeler and E. M. Beesley: "Critique of the Time-Stratigraphic Concept," *Bulletin of the Geological Society of America*, Vol. 59, 1948, p. 84.

As a matter of fact, though it is true that rocks are "dated" by their fossil contents, there are not really even any fossil "unconformities." The faunas of each so-called "age" are now known to grade imperceptibly into those of the next "age." This is true even at the most important boundaries of all — those between the Paleozoic and Mesozoic and between the Mesozoic and Cenozoic.

"A reassessment of the data by Jost Wiedmann of the University of Tubingen in the Federal Republic of Germany gives a clearer picture of evolution at the boundaries of the Mesozoic (225 million to 70 million years ago). He concludes that there were no worldwide extinctions of species or spontaneous appearances of new species at the boundaries. Instead, there was a continuous disappearance of 'old' fauna, and sudden diversification of species that had appeared previously."[1]

It is true that a number of other questions need to be answered before this cataclysmic model will fully explain all the highly-complex and varied geologic phenomena. This is not the place to try to deal with all these questions. Most of them, creationists believe, have been more than adequately resolved in the context of the basic creation model. The usually distinctive fossil assemblages found in different rock systems, for example, are believed to be attributable to a combination of ecologic and hydraulic factors, rather than evolutionary stages. The so-called absolute radiometric dating methods are critically analyzed in terms of their many untenable assumptions and inconsistencies and are likewise found to be assimilable to a concurrent deposition of the bulk of the geologic column. The interested or skeptical reader should acquaint himself with the technical publications of the Institute for Creation Research[2] and of the Creation Research Society before he arbitrarily rejects these new concepts.

---

1 "Fossil Changes: 'Normal Evolution' " A report of papers at the 24th International Geological Congress. *Science News,* Vol. 102, 1972. p. 152.

2 There are a number of books available which attempt to correlate much of the geologic data in the framework of the creation-cataclysm model. Two by the present writer are *The Genesis Flood* (co-author John C. Whitcomb, Jr.),14th printing, 1974, 518 pp., and *Scientific Creationism,* 1974, 277 pp.Both books, as well as many others in this field, are available from the Institute for Creation Research, 2716 Madison Avenue, San Diego, California, 92116.

There is another very important test for comparing the relative effectiveness of the evolution and creation models. We have already discussed the array of organisms in the present world, the nature of biologic changes that occur in the present world, and the character and significance of the fossil record of the past world, in light of the two models. In each case, it has been shown that, while some of these phenomena can possibly be made to correlate with the evolution model, they can also be assimilated more easily and directly within the framework of the creation model. Consequently it is worth raising the question whether or not the basic laws of nature support evolution. Is it really possible for genuine evolution, from a lower to a higher kind, to take place at all?

The Law of Disorder

The evolution model surely must include as its most basic component some "law" of evolution. If indeed all organisms have evolved from primeval random states of matter and energy, there must be in operation some cosmic principle which gradually drives systems from lower degrees to higher degrees of organization and complexity. After all, that is what is supposed to have happened over the astronomic and geologic ages. A man is surely far more complex and highly-ordered than an amoeba, or even than an ape. Whatever this law or principle may be, it is necessary and basic if evolution is to operate. Therefore, it now should be observed operating in systems which are actually evolving. This "law" of increasing order is the most basic and essential prediction of the evolution model.

"I see nothing in modern science to forbid, and much to recommend, the view that evolution stems from the laws built into the very fabric of the universe. Cosmic evolution has given rise to life on at least one minor planet, biological evolution has transcended itself in producing man, and human evolution may in a future perhaps not too remote, be directed by man himself."[1]

The creation model, on the other hand, predicts two quite

---

[1] Theodosius Dobzhansky: "A Biologist's World-View," *Science,* Vol. 175, January 7, 1972, p. 49.

different basic laws in all such systems. Since special creation, by its very nature, implies a specific purpose for each entity specially created (the Creator is not capricious), there will be a fundamental principle in nature which tends to perpetuate the created entities, some principle of *conservation*. Furthermore, any tendency to change these entities cannot result in improvement but rather in detriment to them, since they were created perfect for their purposes in the beginning. Thus, the creation model predicts a universal principle of conservation and another principle, not quite so fundamental but equally universal, of *disintegration*.

Note the clear-cut contrast, which makes it easy to compare the two models. Evolution predicts a universal law of innovation and integration, creation one of conservation and disintegration.

The question is: which fits the observed facts in the real world, most directly and simply? The answer is: the prediction of the creation model is confirmed explicitly by the two laws of thermodynamics; whereas no universal law of innovation and integration has yet been identified at all. This important conclusion is thoroughly documented in Chapter V.

The first law of thermodynamics is essentially the principle of conservation of mass-energy. Everything in the physical universe, so far as we can observe, is being conserved. Energy (including matter) is neither being created nor annihilated, though it can pass through broad changes of form. A similar principle of conservation in the biologic realm seems to insure that, while there are innumerable individual differences within kinds, the basic kinds are permanent (except in cases of extinction). New individuals, varieties, and species may appear, but never new kinds.

The second law of thermodynamics is the law of increasing entropy, stating that all real processes tend to go toward a state of higher probability, which means greater disorder. This law applies to all known systems, both physical and biological, a fact which is universally accepted by scientists in every field.

"As far as we know, all changes are in the direction of in-

creasing entropy, of increasing disorder, of increasing randomness, of running down."[1]

As noted above, this law of increasing entropy is a specific
prediction of the creation model. "Time's arrow," as it has been
called, points downward. The predicted, but hypothetical, "law"
of evolution, points upward. Each is a prediction of change on a
universal scale, but the direction which such changes take
provides the most definitive test of the two models. Each is the
converse of the other.

It may of course be possible to harmonize evolution and entropy
by imposing special conditions on the systems affected. But this
is the very point! One does *not* need to impose special conditions
in order to accommodate the entropy law to the creation model.
Entropy is an explicit *prediction* of creationism!

And what about the conditions that must be satisfied if entropy
is to be accommodated to evolution? Are these possible and
reasonable conditions?

Entropy tends to propel systems from order to disorder. How,
in the face of this pressure, can evolution elevate the same
systems from disorder to order? Why do evolutionists for the
most part ignore this problem?

Those few evolutionists who discuss this problem normally dismiss it with the statement that, since the second law applies only
to closed systems, the energy from the sun acting on the earth as
an open system is sufficient to energize evolution.

But such a naive dismissal is unworthy of scientists. It is, indeed, a vacuous statement, since it contains no information. Certainly the earth is an open system, accessible to the sun's energy.
But so is every other system!

There is no such thing as an isolated system, in the real world.
All systems on the earth are open, directly or indirectly, to the
sun's energy, yet only a very few special systems exhibit a
natural growth in order. If solar energy acting on an open system
were all that is necessary to produce growth, then a pile of bricks
and lumber on a construction site could spontaneously come

---

[1] Isaac Asimov: "Can Decreasing Entropy Exist in the Universe?" *Science
Digest,* May 1973, p. 76.

together to generate an office building. Simpson's comparison is illuminating:

"We have repeatedly emphasized the fundamental problems posed for the biologist by the fact of life's complex organization. We have seen that organization requires work for its maintenance and that the universal quest for food is in part to provide the energy needed for this work. But the simple expenditure of energy is not sufficient to develop and maintain order. A bull in a china shop performs work, but he neither creates nor maintains organization. The work needed is *particular* work; it must follow specifications; it requires information on how to proceed."[1]

If this requirement is necessary to enable an individual organism to grow, requiring the infinitely complex informational system structured into the genetic code for that organism, how much more essential is some sort of a pre-programmed evolutionary code, providing evolution information on how to proceed all the way upward and onward from particle to people — from blob to brain!

Furthermore, in all growth processes, there is present some form of motor for energy conversion. The sun's energy must be converted via *photosynthesis*, for example, into the growth in structure of the plant. But where is the remarkable cosmic converter which transforms solar energy into evolutionary growth? Having energy and an open system is irrelevant — *all* systems have that. The question is *how*?

So far, evolutionists have no answer to such questions. Mutation is not a *code*! It is a random process, and programs nothing. No geneticist has yet demonstrated experimentally or analyzed mathematically how a random mutation can elevate a complex system from a lower degree of order to a higher degree of order. Neither can the passive screen of natural selection accomplish such miracles.

Furthermore, natural selection converts no energy. Mutations only respond to the disintegrative effect of solar energy (as the

---

1 George Gaylord Simpson and W. S. Beck: *Life — An Introduction to Biology* (2nd Ed., New York, Harcourt, Brace and World, Inc., 1965), p. 466.

china to the bull!) but they certainly do not organize that energy into a more complex form of their own molecular systems.

Neither mutation nor selection nor both together (nor any other mechanisms) have yet been demonstrated by evolutionists to provide the necessary energy converting and programming complex which is absolutely required to offset entropy if evolution is to occur in an upward direction. This problem is so critical to the credibility of the evolution model that it will be discussed in much more detail in the next chapter, where both the scientific and theological implications of the entropy principle will be explored in depth.

This objection does not preclude the possibility of evolution. It is just that the necessary "law" of evolution, if it exists, still remains to be discovered and evolutionists must in the meantime continue to exercise faith in their model in spite of entropy.

The creationist does not have this problem. His model fits all the facts of the real world, including entropy, directly and simply, without the necessity for all the secondary assumptions required by evolution. The one difficulty with the creation model is that it implies a Creator, and this suggests that man is responsible to Him. To evolutionists, this is enough to make creationism incredible. But this is a spiritual decision, not a scientific decision.

The Biblical Model of Creation

Up to this point in this chapter, we have considered only the scientific aspects of the two models of origins. It has been shown that, even when considered entirely apart from Biblical revelation on the subject, the actual facts of science can be assimilated much more directly and easily to the creation model than to the evolution model. The latter can be made to fit the data only by the expedient of adding many secondary assumptions and modifications to the basic model, whereas the creation model directly *predicts* these same data.

But since a final decision as to which model to accept is really a spiritual decision, requiring the exercise of faith, it is quite appropriate to bring the record of the Book of Genesis into the discussion at this point. Revelation gives further information about creation which could never be acquired strictly by scientific observations. At the same time, there is nothing in science which can be shown to conflict with the Biblical statements on this subject, even when the latter are taken in the most literal, natural way possible.

The evolution model, on the other hand, is found thereby to be still further weakened, since it contradicts so explicitly the testimony of divine revelation on the subject. The evidence that the Bible is, indeed, divinely inspired and infallibly true, is beyond the scope of this book, but is actually overwhelmingly great. [1]

First, however, we must consider a commonly suggested device for accommodating *both* evolution and creation in the same model — namely, theistic evolution. Although this notion is accepted by very few of the real *leaders* of evolutionary thought, it is at least nominally held by many of their *followers*. Many Christians, who insist they accept the Bible as God's revelation, nevertheless feel they can reinterpret Genesis to make it teach evolution rather than creation. In fact, by this device, they view evolution as God's "method" of creation. The exegetical merits of this interpretation are analyzed more fully in Chapter VI.

It should be recognized, however, that whatever merit the concept of theistic evolution may or may not have as a religious concept, it is *not a scientific model of origins at all.* The idea that some kind of God, or other supernatural agency, may be somehow operating the evolutionary process back behind the scenes may be a belief held on faith, but there is no conceivable way in which it could be evaluated scientifically.

Evolution, as such, can clearly be postulated as a scientific model, as described in preceding sections, and can be used as a framework within which to attempt to correlate and predict scientific data. The same is true of the creation model, as also shown, but there is no way by which an atheistic evolutionary model may be distinguished *scientifically* from a theistic evolutionary model. That is, both concepts are exactly the same insofar as the use of present processes to explain the past is concerned. Both involve the same long chronology and the same sequences of events in the past. Whether the processes operating

---

1 See, for example, *Many Infallible Proofs* by Henry M. Morris (San Diego, Ca., Creation-Life Publishers, 1974), 381 pp.

during these ages, which are supposedly the same processes operating at present, were powered by a divine agency or not can in no way be determined scientifically.

Thus, theistic evolution is strictly and solely a religious model, not a scientific model. It must be evaluated on the basis of religious criteria, not scientific criteria. If evolution is really the best model for explaining the universe, then no God is needed anyway, so far as science is concerned. As Julian Huxley has said:

"Darwin pointed out that no supernatural designer was needed; since natural selection could account for any known form of life, there was no room for a supernatural agency in its evolution."[1]

Or, as Francisco Ayala, one of the younger leaders of evolutionary thought in the world today, has put it:

"Darwin substituted a scientific teleology for a theological one. The teleology of nature could now be explained, at least in principle, as the result of natural laws manifested in natural processes, without recourse to an external Creator or to spiritual or non-material forces."[2]

There are many evangelicals today who, either openly or surreptitiously, believe in theistic evolution. This permits them to maintain a spiritual posture when among other Christians, and also to appear quite up to date and scientific when among unbelievers. However, if they choose to straddle the cosmological fence in this way, they should realize that they are maintaining neither a Biblical nor a scientific position. The evolutionary scientist quite properly insists that, if the evolutionary model is true at all, there is no need for God.

The creationist, on the other hand, frankly acknowledges that God is absolutely necessary and that the present processes of nature cannot possibly explain the origin and development of the world. He maintains that the study of present nuclear and

---

[1] Julian Huxley: *Issues in Evolution,* Ed. by Sol Tax (University of Chicago Press, 1960), p. 45.
[2] Francisco J. Ayala: "Teleological Explanations in Evolutionary Biology," *Philosophy of Science,* Vol. 37, March 1970, p. 2.

astronomic phenomena can *never* explain the origin of the elements or the galaxies or the planets. The study of biochemistry and terrestrial environments can never explain the origin of life, nor can the biologic processes of mutation and natural selection ever account for the origin of the various kinds of organisms.

The creation model therefore postulates a primeval period of *special* creation in the past — "special" in the sense that the processes of creation which were then in operation are no longer in operation today, and are therefore not accessible for scientific measurement and study.

Evolutionists, of course, regard the creation model almost with horror, considering it scientific treason to think that the methods of naturalistic science are forever deprived of access to the processes of creation. A typical evolutionist reaction, for example, is that of Watson:

". . . the theory of evolution itself (is) a theory universally accepted not because it can be proved by logically coherent evidence to be true but because the only alternative, special creation, is clearly incredible."[1]

George Wald similarly prefers to believe in spontaneous generation even though he regards it as impossible scientifically, because, as he says:

". . . the only alternative to some form of spontaneous generation is a belief in supernatural creation. . . ."[2]

But why should it be thought so incredible that God could create? If there is an omnipotent God, then He is surely powerful enough to create, and to do it without the necessity of using preexistent materials or long ages of geologic time. The whole question is simply whether or not there is a God at all.

Obviously no one can *prove* there is *no God*! He may not believe in God, but that is not the same as being sure there *is* none. If the scientific law of cause and effect means anything at

1 D. M. S. Watson: "Adaptation," *Nature,* Vol. 124, 1929, p. 233.

2 George Wald: "Innovation in Biology," *Scientific American,* Vol. 199, September 1958, p. 100.

all (and all scientists use it), then one must recognize that an intelligible universe suggests an Intelligence that caused the universe, and that individual personalities in the universe imply that their First Cause must be a Person.

Nothing but arrogance could insist that the existence of God has been disproved. If there exists even the possibility of a Creator, then the creation model is at least a legitimate option for consideration, and for careful comparison with the evolution model. Even such a convinced evolutionist as Thomas Huxley acknowledged this much.

"... 'creation,' in the ordinary sense of the word is perfectly conceivable. I find no difficulty in conceiving that, at some former period, this universe was not in existence and that it made its appearance in six days (or instantaneously, if that is preferred), in consequence of the volition of some pre-existing Being. Then, as now, the so-called *a priori* arguments against Theism, and, given a Deity, against the possibility of creative acts, appeared to me to be devoid of reasonable foundation."[1]

Creationists are on good ground when they insist that the creation model be accepted on the same basis as the evolution model — that is, as a framework for the correlation and prediction of scientifically observable data. Though neither can be ultimately proved, since they involve non-repeatable historical events, they can both be used as scientific models, and a decision made between them on the basis of which serves most simply and effectively to accommodate and predict data.

In addition to a period of special creation in the past, the creation model also postulates that the present is a period of *conservation* rather than creation. If the Creator created things in the past, He must have had a purpose in so doing, and therefore He is now carrying out that purpose and is accordingly holding things together.

However, the atheist sometimes objects on the basis of his own moral judgment that if indeed God is upholding all things, He seems in many cases to be doing a poor job of it, allowing such anomalies as disease, struggle, decay, death and the like.

---

[1] Thomas H. Huxley, quoted in *Life and Letters of Thomas Henry Huxley* Ed. by Leonard Huxley (Vol. II, Macmillan, 1903), p. 429.

Either, he says, God is not really good to allow such things, or else He is unable to stop them, and thus is not really God.

This "problem of pain" may constitute a theological problem, with which theists must deal on theological grounds, but it really is no problem from a scientific point of view. The creation model, in order to accommodate it, merely includes in its system another principle operating in the present in addition to the principle of conservation; namely, a principle of deterioration from the original created perfection.

Though not necessarily implicit in the basic creation model, there is still another factor which is normally included, in view of the great evidences found in the earth of primeval catastrophe. Not only was there a period of special creative processes in the past, but also one or more periods of special catastrophic destructive processes. Again, this concept may entail theological problems, but at this point we are considering it only as a part of our model of scientific creationism.

To repeat and summarize, the scientific creation model includes the following basic components:

(1) An initial period of special creation, in which all the basic entities (astronomical systems, structure of matter and energy, the forces and laws of nature, the distinctive kinds of plants and animals, and man) were brought into existence by processes of supernatural creation which no longer operate in the natural world.

(2) The present period in which the basic laws and processes of nature are essentially uniform. In particular, these processes are functioning within a framework of quantitative conservation and qualitative deterioration. Nothing is being created, since this was finished in the special creation period in the past, but neither is anything passing out of existence. And yet, though nothing is lost, things do appear to be running down. Individual organisms, and even entire species, may die out, but life goes on, and always in the same basic distinctive forms as originally created.

(3) One or more periods of cataclysm in the past, when the entire world was affected by destructive processes of tremendous scope and intensity.

The previous statement of the nature of the creation model is rather complete and specific, and certainly is capable of evaluation in terms of the observed data in the real physical world. It is not, as many evolutionists claim, merely a religious belief. It deals with actual systems and processes in the world of observation and experience. It will either fit those data or it will not fit those data, but at least it ought not to be rejected until such an examination of the data has been made.

Thus the evolution and creation models provide two clear-cut alternative ways in which to compare all the data of the real world. The evolution model attempts to explain all things in terms of present processes. The creation model says that periods of special creation and cataclysm in the past are necessary to explain the data. The question is simply which of the two models serves best in the prediction and correlation of the phenomena and data of the world of experience as studied by science.

As seen earlier in this chapter, the creation model does, indeed, fit the actual observed data of science at least as well as — in fact, much better than — the evolution model.

Now, although the basic creation model offers a clear alternative to the evolution model, it can be made more specific by framing it around the Biblical revelation of earth pre-history. It will be found that this framework, when simply and consistently applied, also wonderfully coordinates and assimilates all known data of natural science.

According to the Bible, the period of special creation occupied six days, following which the Creator "rested from all His work which God created and made" (Genesis 2:3). Since that time, He has been "upholding all things" (Hebrews 1:3). The present cosmos is being "kept in store" (II Peter 3:7) under the domain of *conservation.*

However, because of man's sin, God has placed a Curse "on the ground" (Genesis 3:17), so that the very "dust of the earth" — the fundamental elements out of which all things had been constructed by God — has been under the "bondage of decay" (Romans 8:21). Thus it is that "the whole creation groaneth and travaileth in pain together until now" (Romans 8:22). Everything, therefore, is heading downhill toward decay and death. The world is thus under the domain of *disintegration* as well as *conservation*.

These two universal principles, so clearly set forth in the Bible, of course are now formally incorporated in the basic structure of modern science, as we have seen, and have come in the past century to be denominated as the Two Laws of Thermodynamics. The First Law is the Law of Conservation; the Second Law is the Law of Disintegration. The creation is constant in quantity, but deteriorating in quality. The total energy remains unchanged, but the available energy continually decreases.

Another facet of the Biblical model, of course, is the great Flood of the days of Noah. This is described in the Bible as a worldwide cataclysm of unprecedented and unequalled magnitude. "The world that was, being overflowed with water, perished" (II Peter 3:16).

The Flood was obviously primarily a hydraulic, and therefore sedimentary, phenomenon. Torrential waters poured down from the skies and gushed forth from the fountains of the great deep, for 150 days, all over the world. Accompanying these activities must necessarily have been a great complex of atmospheric violence, volcanic eruptions, earth movements, giant waves and other disturbances that profoundly altered the face of the earth. The world that emerged when the waters receded bore little resemblance to the beautiful antediluvian earth as created by God in the beginning.

The key to the innumerable evidences of catastrophism in the earth's crust and on its surface is found in the Flood and its after-effects. This is especially important in the proper interpretation of the record of the fossils in the earth's sedimentary rocks — a record which has mistakenly been appropriated by evolutionists as the chief evidence for their system.

The Biblical framework of pre-history also includes one other event of worldwide significance; namely, the confusion of tongues and dispersion of the nations at the tower of Babel. Although this event had little effect on the physical world, it did of course have profound effects on the world of mankind, and answers many questions in human history to which the evolution model has never provided any solution at all.

The Biblical creation model is specifically outlined in the first eleven chapters of Genesis and is referred to in one way or another frequently throughout the entire Bible. It amplifies and particularizes the basic model of scientific creationism as discussed earlier. By way of summary and review, it centers around four great events of early earth history, each of *worldwide* significance, as follows:

(1) Special creation of all things in six days, by creative and integrative processes which no longer are in operation, following which God "rested."

(2) The curse on all things, by which the entire cosmos was brought into a state of gradual deterioration leading toward death.

(3) The universal Flood, which drastically changed the rates of most earth processes and the structure of the earth's surface.

(4) The dispersion at Babel, which resulted from the sudden proliferation of languages and other cultural distinctives among the nations and tribes of men.

The Biblical creationist maintains that all the real data of observation correlate perfectly with the above model. There are a great many measurable and observable phenomena in nature, and all of them, without exception, fit this Biblical model, evolutionary philosophers to the contrary notwithstanding.

Evolutionists have fostered the strange belief that everything is involved in a process of progress, from chaotic particles billions of years ago all the way up to complex people today. The fact is the most certain laws of science state that the real processes of nature do not make things go uphill, but downhill. Evolution is scientifically impossible!

Ryuzu Falls, Nikko, Japan

Photo credit: Col. M. G. McBee

CHAPTER V

CAN WATER RUN UPHILL?

Definitions

The study of biological processes and phenomena indicates that significant evolutionary developments are not observable in the modern world. Similarly the great gaps in the fossil records make it extremely doubtful that any genuine evolution, as distinct from small changes within the kinds, ever took place in the past.

As discussed briefly in the preceding chapter, however, there is one weakness in evolutionary theory which goes well beyond the implications of the above difficulties. Not only is there no evidence that evolution ever *has* taken place, but there is also firm evidence that evolution never *could* take place. *The Law of Increasing Entropy* is an impenetrable barrier which no evolutionary mechanism yet suggested has ever been able to overcome. Evolution and entropy are opposing and mutually exclusive concepts. If the entropy principle is really a universal law, then evolution must be impossible.

The very terms themselves express contradictory concepts.

The word "evolution" is of course derived from a Latin word meaning "out-rolling." The picture is of an outward-progressing spiral, an unrolling from an infinitesimal beginning through ever-broadening circles, until finally all reality is embraced within.

"Entropy," on the other hand, means literally "in-turning." It is derived from the two Greek words *en* (meaning "in") and *trope* (meaning "turning"). The concept is of something spiralling inward upon itself, exactly the opposite concept to "evolution." Evolution is change outward and upward, entropy is change inward and downward.

The Biblical usage of these terms is interesting and significant. Neither the word nor the concept of evolution is found in the Bible at all. It is evidently an idea completely alien to Scripture, except as associated with the philosophy of idolaters and skeptics (II Peter 3:4; Romans 1:21-25; Jeremiah 2:27, etc.).

The Greek word *entrope,* however, is found in I Corinthians 6:5 and 15:34, and is translated as "shame" in both. It is also found in the Greek Septuagint translation of Psalm 35:26, 44:15, and 109:29, where it is translated "confusion," and in Psalm 69:19 and 71:13, where it is translated "dishonour." The concept involved is that one who is ashamed or confused simply "turns inward," no longer able to face the outside world. Alternatively, it may be understood in the sense that, if one turns inward (for strength or wisdom) instead of to an external source of supply, he will inevitably be brought ultimately to shame or confusion.

Very appropriately, therefore, the law of increasing entropy states that, in any isolated system, the state of disorder must increase; a system which feeds only on itself must eventually run down. There is another interesting verse in which only the root word *trope* appears, translated as "turning." The verse is James 1:17: "Every good gift and every perfect gift is from above, and cometh down from the Father of lights, with whom is no variableness, neither shadow of turning." Perhaps it is reasonable to paraphrase the last part of this verse in the language of modern science somewhat as follows: "(They) descend from the one ultimate first cause of light and all other forms of energy, with whom there can be neither change in the total quantity of His power nor change in its quality."

With God, therefore, is no change. He is eternal and immutable. However, in this present world, everything is under a rule of change. The question is whether the change is up or down, evolution or entropy.

That the principles of evolution and entropy are both believed to be universal principles and yet are mutually contradictory is seen from the following authoritative definitions.

"(There is a) general natural tendency of all observed systems to go from order to disorder, reflecting dissipation of energy available for future transformation — the law of increasing entropy."[1]

As far as evolution is concerned, the classic definition of Sir Julian Huxley is as follows:

"Evolution in the extended sense can be defined as a directional and essentially irreversible process occurring in time, which in its course gives rise to an increase of variety and an increasingly high level of organization in its products. Our present knowledge indeed forces us to the view that the whole of reality *is* evolution — a single process of self-transformation."[2]

Thus, in the one instance, "all observed systems . . . go from order to disorder," and in the other, "the whole of reality . . . gives rise to an increasingly high level of organization in its products." It seems obvious that either evolution or entropy has been vastly overrated or else that something is wrong with the English language.

The entropy principle, however, is nothing less than the Second Law of Thermodynamics, which is as universal and certain a law as exists in science. First, however, before discussing the Second Law, we should define the First Law and, for that matter, thermodynamics itself.

Thermodynamics is a compound of two Greek words, *therme* ("heat") and *dunamis* ("power"). It is the science that treats the power or energy contained in heat, and its conversion to other

---

[1] R. B. Lindsay: "Physics — To What Extent is it Deterministic?" *American Scientist,* Vol. 56, Summer 1968, p. 100.

[2] Julian Huxley: "Evolution and Genetics," in *What is Man?,* Ed. by J. R. Newman (New York, Simon and Schuster, 1955), p. 278.

forms of energy. The term "energy" is itself derived from the Greek word *energeia* ("working"), and is normally defined as "the capacity to do work." In modern scientific terminology, "energy" and "work" are considered equivalent, each measured as the product of a force times the distance through which it acts (foot-pounds, in the English system of dimensions). Something which has "energy" has the "capacity to do work;" that is, the "capacity to exert a force through a distance."

The concept of "power" is closely related to that of "energy," except that the time factor must also be taken into account. Power is the work done, or the energy expended to do the work, per unit of time — measured in foot-pounds per second.

Thus the science of thermodynamics is the study of the conversion of heat power into mechanical power, the use of steam to turn the wheels and move the load. The invention of the steam engine, and the development of the theoretical equations of thermodynamics by which to design it, led to the great Industrial Revolution and our modern age of technology. The golden age of science and the greatest scientists of all — Newton, Maxwell, Kelvin, and others — produced the discipline and the Laws of Thermodynamics.

It was not long after the discovery of the "mechanical equivalent of heat" by Joule and others before it was realized that there were also other forms of energy (electrical energy, chemical energy, light, heat, sound, etc.) and that all of them were comprehended within these same Laws of Thermodynamics The advent of the atomic age made it evident that even matter itself was merely another form of energy, and that it also could be brought under the broad umbrella of thermodynamics.

Thus the modern scientist has come to recognize that the science of thermodynamics is exceedingly broad. It provides the basic framework for all energy conversion processes, whereby heat, electricity, or any other form of energy can be converted into any other form. In the twentieth century, it also includes the processes of mass-energy interchanges.

The First Law of Thermodynamics

Since all processes are fundamentally energy conversion

processes, and since everything that happens in the physical universe is a "process" of some kind, it is obvious why the Two Laws of Thermodynamics are recognized as the most universal and fundamental of all scientific laws. Everything that *exists* in the universe is some form of energy, and everything that *happens* is some form of energy conversion. Thus the Laws which govern energy and energy conversion are of paramount importance in understanding the world in which we live.

It should be stressed that these Laws are empirical laws. That is, like all other laws of science, they are accepted on the basis of experience and testing, not because of some deterministic mathematical proof. However, they are based on better and more varied evidence than any other scientific principles whatever.

Isaac Asimov defines the First Law as follows:

"To express all this, we can say: 'Energy can be transferred from one place to another, or transformed from one form to another, but it can be neither created nor destroyed.' Or we can put it another way: 'The total quantity of energy in the universe is constant.' When the total quantity of something does not change, we say that it is conserved. The two statements given above, then, are two ways of expressing 'the law of conservation of energy.' This law is considered the most powerful and most fundamental generalization about the universe that scientists have ever been able to make."[1]

As long as nuclear reactions are not involved, the Law of Conservation of Matter also applies. However, matter can be converted into energy and energy into matter, and in either case, the sum total of matter and energy remains the same. Similarly, as long as the process involves only changes in kinetic energy (that is, energy of motion), then the Law of Conservation of Momentum applies. However, the most comprehensive of all the conservation laws is the Law of Conservation of Mass-Energy, and this is the First Law of Thermodynamics.

Asimov makes a very interesting point when he says concerning this Law:

---

1 Isaac Asimov: "In the Game of Energy and Thermodynamics You Can't Even Break Even," *Smithsonian Institute Journal,* June 1970, p. 6.

"No one knows *why* energy is conserved."[1]

He should have said, of course, that *science* cannot tell us why energy is neither created nor destroyed. The Bible, however, does give us this information.

The reason why no energy can now be created is because only God can create energy and because God has "rested from all His work which He created and made" (Genesis 2:3). The reason why energy cannot now be destroyed is because He is now "upholding all things by the word of His power" (Hebrews 1:3). "I know that, whatsoever God doeth, it shall be forever: nothing can be put to it, nor anything taken from it" (Ecclesiastes 3:14).

The Second Law in Classical Thermodynamics

The First Law is itself a strong witness against evolution, since it implies a basic condition of stability in the universe. The fundamental structure of the cosmos is one of conservation, not innovation. However, this fact in itself is not impressive to the evolutionist, as he merely assumes that the process of evolution takes place within the framework of energy conservation, never stopping to wonder where all the energy came from in the first place nor how it came to pass that the total energy was constant from then on.

It is the Second Law, however, that wipes out the theory of evolution. There *is* a universal process of change, and it *is* a directional change, but it is *not* an upward change.

This entropy law appears in three main forms, corresponding to classical thermodynamics, statistical thermodynamics, and informational thermodynamics, respectively. Each of these corresponds to a different, though equivalent, concept of entropy.

In so-called classical thermodynamics, the Second Law, like the First, is formulated in terms of energy.

"It is in the transformation process that Nature appears to exact a penalty and this is where the second principle makes its appearance. For every naturally occurring transformation of energy is accompanied, somewhere, by a loss in the *availability* of energy for the future performance of work."[2]

---

[1] *Ibid.*

[2] R. B. Lindsay: "Entropy Consumption and Values in Physical Science," *American Scientist,* Vol. 47, September 1959, p. 378.

In this case, entropy can be expressed mathematically in terms of the total irreversible flow of heat. It expresses qualitatively the amounts of energy in an energy conversion process which becomes unavailable for further work. In order for work to be done, the available energy has to "flow" from a higher level to a lower level. When it reaches the lower level, the energy is still in existence, but no longer capable of doing work. Heat will naturally flow from a hot body to a cold body, but not from a cold body to a hot body.

For this reason, no process can be 100% efficient, with all of the available energy converted into work. Some must be deployed to overcome friction and will be degraded to non-recoverable heat energy, which will finally be radiated into space and dispersed. For the same reason a self-contained perpetual motion machine is an impossibility.

Since, as we have noted, everything in the physical universe is energy in some form and, since in every process some energy becomes unavailable, it is obvious that ultimately, *all* energy in the universe will be unavailable energy, if present processes go on long enough. When that happens, presumably all the various forms of energy in the universe will have been gradually converted through a multiplicity of processes into uniformly (that is, randomly) dispersed heat energy. Everything will be at the same low temperature. There will be no "differential" of energy levels, therefore no "gradient" of energy to induce its flow. No more work can be done and the universe will reach what the physicists call its ultimate "heat death."

This is a depressing outlook for the future, but is what must come eventually, if present processes continue. The fact that the universe has not yet reached this dead condition, and in fact is still very much alive, with tremendous reservoirs of available energy everywhere, proves that it is not infinitely old. If it were of infinite antiquity, it obviously would already be dead.

Thus the Second Law proves, *as certainly as science can prove anything whatever,* that the universe had a beginning. Similarly the First Law shows that the universe could not have begun itself. The total quantity of energy in the universe is constant, but the quantity of *available* energy is decreasing.

Therefore, as we go *backward* in time, the available energy would have been progressively greater until, finally, we would reach the beginning point, where available energy equalled total energy. Time could go back no farther than this. At this point both energy and time must have come into existence. Since energy could not create itself, the most scientific and logical conclusion to which we could possibly come is that: "In the beginning, God created the heaven and the earth."

The evolutionist will not accept this conclusion, however. He hypothesizes that either: (1) some natural law cancelling out the Second Law prevailed far back in time, or (2) some natural law cancelling out the Second Law prevails far out in space.

When he makes such assumptions, however, he is denying his own theory, which says that all things can be explained in terms of presently observable laws and processes. He is really resorting to creationism, but refusing to acknowledge a Creator.

Entropy and Disorder

A second way of stating the entropy law is in terms of statistical thermodynamics. It is recognized today not only that all scientific laws are empirical but also that they are statistical. A great number of individual molecules, in a gas for example, may behave in such a way that the overall aspects of that gas produce predictable patterns in the aggregate, even though individual molecules may deviate from the norm. Laws describing such behavior must be formulated statistically, or probabilistically, rather than strictly dynamically. The dynamical laws then can theoretically be deduced as limiting cases of the probabilistic statements.

In this context, entropy is a probability function related to the degree of disorder in a system. The more disordered a system may be, the more likely it is.

"All real processes go with an increase of entropy. The entropy also measures the randomness, or lack of orderliness of the system; the greater the randomness, the greater the entropy."[1]

Note again the universality expressed here — *all real*

---

[1] Harold Blum: "Perspectives in Evolution," *American Scientist,* October 1955, p. 595.

processes. Isaac Asimov expresses this concept interestingly as follows:

"Another way of stating the Second Law then is: 'The universe is constantly getting more disorderly.' Viewed that way, we can see the Second Law all about us. We have to work hard to straighten a room, but left to itself, it becomes a mess again very quickly and very easily. Even if we never enter it, it becomes dusty and musty. How difficult to maintain houses, and machinery, and our own bodies in perfect working order; how easy to let them deteriorate. In fact, all we have to do is nothing, and everything deteriorates, collapses, breaks down, wears out, all by itself — and that is what the Second Law is all about." [1]

Remember that this tendency from order to disorder applies to all real processes. Real processes include, of course, biological and geological processes, as well as chemical and physical processes. The interesting question is: "How does a real biological process, which goes from order to disorder, result in evolution, which goes from disorder to order?" Perhaps the evolutionist can ultimately find an answer to this question, but he at least should not ignore it, as most evolutionists do.

Especially is such a question vital, when we are thinking of evolution as a growth process on the grand scale from atom to Adam and from particle to people. This represents an absolutely *gigantic* increase in order and complexity, and is clearly out of place altogether in the context of the Second Law.

Information Theory and the Second Law

A third and still more fascinating concept of entropy comes from the field of information theory, or what may be called communicational thermodynamics. This new scientific discipline is a product of the computer age and the field of cybernetics. Informational science attempts to quantify the communication of meaningful information from sender to receiver. Communications systems include books, television sets, tape recorders, computer tapes, and many other devices. Even a man and his brain can be considered as such a system.

---

[1] Isaac Asimov: "In the Game of Energy and Thermodynamics You Can't Even Break Even," *Smithsonian Institute Journal,* June 1970, p. 6.

In information theory, entropy is considered to be a measure of the degree in which information is lost or becomes garbled in the transmission process. It measures the "noise" or "static" which tends to inhibit the perfect transmission of a message. The process of communication is surprisingly analogous to a standard energy conversion process. Just as some energy is lost in the conversion process, so always some information is lost in the communication process. No one appropriates the pastor's complete sermon, and the recording never reproduces the orchestral rendition with perfect fidelity.

One can sense intuitively that all three concepts of entropy are similar and basically equivalent to each other. As a matter of fact, this equivalence can be demonstrated quite rigorously by mathematics, though the proof is much too complicated to try to reproduce here. The fact that entropy is really the same entity, whether defined in classical, statistical, or informational terms, is in fact one of the major discoveries of modern science.

The equivalence of entropy in the classical and statistical contexts is implied in the following:

"Each quantity of energy has a characteristic quality called entropy associated with it. The entropy measures the degree of disorder associated with the energy. Energy must always flow in such a direction that the entropy increases."[1]

Similarly, the equivalence of these concepts with the informational concept is recognized.

"It is certain that the conceptual connection between information and the Second Law of Thermodynamics is now firmly established."[2]

All sorts of intriguing illustrations of these relationships may be discovered. For example, the sun's energy is converted through photosynthesis into vegetables. These are eaten by a man, whose metabolic processes convert their stored chemical energy into energy-imparting molecules transmitted through the blood stream to various parts of the body, especially the brain.

[1] Freeman J. Dyson: "Energy in the Universe," *Scientific American,* Vol. 224, September 1971, p. 52.

[2] Myron Tribus and Edward C. McIrvine: "Energy and Information," *Scientific American,* Vol. 224, September 1971, p. 188.

The blood energizes the complex brain cells and circuitry, which then generate thought and convey information.

Isaac Asimov confirms that all these different ways of looking at the Second Law are really equivalent to each other.

"That is one way (that is, decreasing availability of energy — author) of stating what is called the Second Law of Thermodynamics. It is one of many ways; all of them are equivalent although some very sophisticated mathematics and physics is involved in showing the equivalence."[1]

Thus, there are three basic vehicles of physical reality associated with the entropy concept. In the structure of all systems, entropy is a measure of *disorder*. In the maintenance of all processes, entropy is a measure of *wasted energy*. In the transmission of all information, entropy is a measure of *useless noise*. Each of these three concepts is basically equivalent to the other two, even though it expresses a distinct concept.

Always, furthermore, entropy tends to increase. Everywhere in the physical universe there is an inexorable downhill trend toward ultimate complete randomness, utter meaninglessness, and absolute stillness. In the beginning was the Father of creation, the Word of communication, and the Spirit of power. In the end is nothing but disorder, confusion and death. The evolutionary delusion becomes absolute nonsense in the context of the all-comprehensive Second Law.

Are There Exceptions to the Second Law?

When confronted with the anti-evolutionary implications of the Second Law of Thermodynamics, evolutionists exhibit a variety of reactions. Most of them appear to be ignorant of the fact that there is such a thing. Even those that have heard about it for the most part have never thought about its possible implications for their theory.[2] It is surprising, but true, that most textbooks dealing with some aspect of organic evolution never even mention the entropy principle.

---

[1] Isaac Asimov: *op cit,* p. 8.

[2] The writer and his creationist colleagues have participated in many formal debates with prominent evolutionists on university campuses. Never have any of them been able to deal effectively with the entropy question. Most of them ignore it, even when their opponents have made a major issue of it in the debate.

Some evolutionists take refuge in the idea that, since the universe is almost infinitely large, and we can only sample a small part of it, we don't really know that the entropy principle always applies. However, what we *do* know is that, wherever it has been tested, it always works. Whether or not the Second Law may hold on some hypothetical planet a million light years away has no bearing on the fact that it always holds true on *this* planet, where evolution is supposed to be happening.

Others suppose that, since entropy is "statistical," there may be occasional exceptions to the entropy Law, and that evolution on the earth is such an exception. Again, however, it is on the earth that science has *proved* the Law! Even if a rare accidental spurt in some process violated entropy and created an evolutionary gain of some sort, the next spurt would undoubtedly be in the opposite direction and undo it.

Most knowledgeable evolutionists, however, if pushed for an answer to the entropy problem, will take refuge in the "open system" argument. Asimov says:

"Life on earth has steadily grown more complex, more versatile, more elaborate, more orderly, over the billions of years of the planet's existence. . . . How could that vast increase in order (and therefore that vast decrease in entropy) have taken place? The answer is it could *not* have taken place without a tremendous source of energy constantly bathing the earth, for it is on that energy that life subsists. . . . In the billions of years that it took for the human brain to develop, the increase in entropy that took place in the sun was far greater, — far, far greater than the decrease that is represented by the evolution required to develop the human brain." [1]

In other words, the earth in its geologic time setting is "open" to the sun's energy, and it is this tremendous influx of energy which powers the evolutionary process and enables it to rise and overcome the entropy law which would otherwise inhibit it. The First and Second Laws of Thermodynamics, they say, apply only to *isolated systems* — systems into which no external energy can flow — and so supposedly do not apply to the earth.

---

[1] Asimov: *op. cit.,* p. 11.

The evolutionist will also cite various examples of growth in open systems to illustrate his point, such as a seed growing up into a tree with many seeds. In like manner, he says, the sun supplies energy to the open earth-system throughout geologic time to keep evolution going, even though perhaps at some long-distant time the greater earth-sun system will finally die and evolution will stop.

This is an exceedingly naive argument and it indicates the desperate state of evolutionary theory that leads otherwise competent scientists to resort to it. It should be self-evident that the mere existence of an open system of some kind, with access to the sun's energy, does not of itself generate growth. The sun's energy may bathe the site of an automobile junk yard for a million years, but it will never cause the rusted, broken parts to grow together again into a functioning automobile. A beaker containing a fluid mixture of hydrochloric acid, water, salt, or any other combination of chemicals, may lie exposed to the sun for endless years, but the chemicals will never combine into a living bacterium or any other self-replicating organism. More likely, it would destroy any organisms which might accidentally have been caught in it. Availability of energy (by the First Law of Thermodynamics) has in itself no mechanism for thwarting the basic decay principle enunciated by the Second Law of Thermodynamics. *Quantity* of energy is not the question, but *quality!*

Criteria for a Growth Process

The Second Law says that all processes basically must be decay processes. Apparent exceptions to this rule do exist, however, especially in the phenomena of life. A seed grows up into a tree, and an embryo grows up into an adult animal. Even in the non-living world, there seem to be some exceptions; for example, the formation and growth of a crystal. And of course there are many artificial growth processes which can be produced. Threads can be made to grow into a dress and bricks can be made to grow into a building.

It should never be forgotten, however, that all such apparent decreases of entropy can only be produced at the expense of a still greater increase of entropy in the external environment, so that the world as a whole, or the universe as a whole, continues to run down. Furthermore, such growth processes are only temporary at best. The tree and the animal eventually die and the crystal sooner or later disintegrates. The dress wears out and the building crumbles. "Dust to dust" is always the victor in the end.

We need yet to consider the problem of the production of even a local, temporary growth process. What criteria have to be satisfied before such a "negentropic" process becomes possible, and does the supposed evolutionary process meet these criteria?

A little consideration quickly makes it evident that at least *four* criteria have to be satisfied before a growth process can be initiated and maintained. The first two of these are obvious:

(1) *An open system.* Obviously growth cannot occur in a closed system; the Second Law is in fact *defined* in terms of a closed system. However, this criterion is really redundant, because in the real world closed systems do not even exist! It is obvious that the Laws of Thermodynamics apply to open systems as well, since they have only been tested and proved on open systems!

(2) *Available Energy.* This criterion is also actually redundant, since the energy of the sun is always available, either directly or indirectly, to all systems of any kind on the entire earth. As the Scripture says, "There is *nothing* hid from the heat thereof" (Psalm 19:6).

Now, however, we come to the real heart of the problem. The evolutionist glibly gives entropy the brush-off because the earth is an open system bathed in the sun's energy. Such an answer is vacuous and trivial, since *all* systems are open to the sun's energy, but only a few exhibit a growth process, and even these only temporarily. What must be the remarkable additional conditions that can empower a worldwide evolutionary growth process in the whole biosphere for three billion years?

For even the local, temporary growth systems with which men have observational acquaintance (as distinct from philosophical predilection), there must be at least two additional criteria satisfied.

(3) *A Coded Plan.* There must always, without known

124

exception, exist a pre-planned program, or pattern, or template, or code, if growth is to take place. Disorder will never randomly become order. Something must sift and sort and direct the environmental energy before it can "know" how to organize the unorganized components. The fact that a "need" exists for growth to take place is of little moment to bobbing particles.

In the case of the plant, for example, the necessary program for its growth has been written into the structure of the germ cells, including especially the *genetic code*, the amazing system of the DNA-RNA complex which somehow, by its intricately-coiled template structure and "messenger" functions, directs the assimilation of the environmental chemicals into a resulting plant structure like that of its predecessor plants. A similar coding system is also present in the animal seed. Note the statement by Simpson and Beck (see p. 100).

In inorganic systems, the growth is directed by the intricate molecular structure of the crystal compound and by the chemical properties of the elements comprising it. Each crystal is directed into a predictable geometric pattern on the basis of the chemical code implicit in the periodic table of the elements and their own pre-existing structures.[1]

Artificial processes also have their "codes." The building is based on a blueprint and the dress on a pattern.

But whence came these codes? How did the chemical elements acquire their orderly properties? What primeval DNA molecule had no previous DNA molecule to go by?

Our experience with artificial processes indicates that a code for growth requires an intelligent planner. An architect had to draw the blueprint and a dress designer prepared the pattern. Could mindless, darting particles plan the systematic structure of the elements that they were to form? Even more unbelievably, could these elements later get together and program the genetic

---

[1] As a matter of fact, in the case of the crystallization process, there is an important sense in which the formation of a crystal really represents an *increase* of entropy, even in the solution from which it forms. That is, there is more "information" or "energy," for the production of work, in the liquid solution than in the stable structure which crystallizes out of it.

code, which could not only direct the formation of complex living systems of all kinds, but even enter into the replication process which would insure the continued production of new representatives of each kind? To imagine such marvels as this is to believe in magic — and magic without even a magician at that!

A code *always* requires an intelligent coder. A program requires a programmer. To say that the most fantastically complex and effective code of all — the genetic code — somehow coded itself in the first place, is to abandon all pretence of science and reason in the study of the world as it is.

But the genetic code is utter chaos in comparison with the complexity of a program which might conceivably direct the evolutionary growth process from particles to people over five billion years of earth history! Where is the evidence for such a program? What structure does it have? How does it function, in order to direct elements into proteins and proteins into cells, cells into plants and invertebrates, fishes into birds, and monkeys into men?

The sun's energy is there all right, and the earth is assuredly an open system, but by what marvelous automated directional system is this energy instructed how to transmute a school of jellyfish into a colony of beavers?

Does the evolutionist imagine that mutation and natural selection could really perform the function of such an unimaginably complex program? Mutation is not a code, but only a random process which, like all random processes, generates disorder in its products. Natural selection is not a code, but only a sort of cybernetic device which snuffs out the disorderly effects of the mutation process. Is the evolutionist really so foolish as to think this kind of mindless interplay could produce the human brain — or, is it not simply that "the god of this world hath blinded the minds of them who believe not"? (II Corinthians 4:4).

But there is still another criterion which must be satisfied, even for a local temporary growth process:

(4) *An Energy-Conversion Mechanism.* It is naively simplistic merely to say: "The sun's energy sustains the evolutionary process." The question is: "*How* does the sun's energy sustain the evolutionary process?" This type

of reasoning is inexcusable for scientists, because it confuses the First Law of Thermodynamics with the Second Law. There is no doubt that there is a large enough *quantity* of energy (First Law) to support evolution, but there is nothing in the simple heat energy of the sun of sufficiently high *quality* (Second Law) to produce the infinitely-ordered products of the age-long process of evolutionary growth.

One could much more reasonably assume that the sun's energy bathing the stockpiles of bricks and lumber on a construction site will by itself erect an apartment building, an infinitely simpler structural project than the supposed products of organic evolution. There is far more than enough energy reaching the building site than is necessary to build the building, so why bother to rent equipment and hire workmen? This very reasonable suggestion will not work, however, even if the sun's heat bears down on those materials for a billion years.

The missing ingredient is an energy-conversion mechanism! Some mechanism has to be on hand to convert the sun's energy into the mechanical energy required to erect the structure. This is *always* true, for any growth process. The natural tendency is to decay, so that for growth to take place, some very special and effective mechanism must be superimposed to convert the simple heat energy into the complex growth system.

In the case of the seed growing up into a tree, for example, the mechanism is that of photosynthesis. This is a marvelous and intricate mechanism by which the sun's radiant energy is somehow transformed into the growing plant tissue. Photosynthesis is so complex and wonderful a mechanism that scientists even yet do not fully comprehend it,[1] involving as it does an involved combination of electrochemical reactions, bacterial agencies, and other factors.

Similarly, various metabolic mechanisms convert the chemical energy stored in the plant into the mechanical and other forms of energy which the animal that eats the plant needs in his activities. The plant's energy may also eventually be converted into

[1] R. P. Levine: "The Mechanism of Photosynthesis," *Scientific American,* Vol. 221, December 1969, pp. 58-70. Mr. Levine is Professor of Biology at Harvard.

coal, the burning of which may drive a boiler which produces steam for a generator to make electrical energy. The latter is available at the construction site for conversion into the mechanical energy necessary for the construction equipment as it is operated to build the building.

Dr. Lewis Thomas expresses this requirement in somewhat different terminology, using the concept of the "membrane," especially regarding the earth's atmosphere-biosphere complex as such a structure.

"It takes a membrane to make sense out of disorder in biology. You have to be able to catch energy and hold it, storing precisely the needed amount and releasing it in measured shares. A cell does this, and so do the organelles inside. Each unit is poised in the flow of solar energy, tapping off energy from metabolic surrogates of the sun. To stay alive, you have to be able to hold out against equilibrium, maintain imbalance, bank against entropy. In our kind of world, you can only transact this business with membranes."[1]

But what is the source of such marvelously designed systems which thus maintain life and permit its reproduction in spite of entropy? Thomas says:

"You could say that the breathing of oxygen into the atmosphere was the result of evolution, or you could turn it around and say that evolution was the result of oxygen. You can have it either way."[2]

Which is one convenient way of avoiding hard questions! Evolution must indeed be a marvelous phenomenon if it can evolve the conditions necessary for its own origin. But that is not all.

"It is another illustration of our fantastic luck that oxygen filters out the very bands of ultraviolet light that are most devastating to nucleic acids and proteins, while allowing full penetration of the visible light needed for photosynthesis. If it had not been for this semipermeability, we could never have come along."[3]

---

1 Lewis Thomas: "The Miraculous Membrane," *Intellectual Digest,* Vol. IV, February 1974, p. 56.
2 *Ibid.*
3 *Ibid.*

We are, indeed, very lucky! Had it not been for this fantastic mechanism that "accidentally" developed, the law of entropy would never have allowed living things to exist at all.

Always, therefore, one or more energy conversion mechanisms must be available for utilization of the sun's energy whenever there is any kind of growth process. This is in addition to the pre-programmed plan for directing the growth process, which must also be available.

But the most extensive and energy-demanding growth process of all — namely, the organic evolution of the entire biosphere — has no such mechanism! Neither does it have, as we have seen, a program. How, then, can it possibly work?

The evolutionists' answer, of course, is: *mutation and natural selection*. However, neither of these constitutes a coded plan. Likewise, neither of them is an energy conversion mechanism. If neither is either, how can both be both?

A mutation is a random change in an already ordered code, and natural selection is a sort of sieve which screens out these mutants and thus preserves this previous order. Neither one has the potential to convert the sun's energy (or any other kind of energy) into some other form of energy which will generate new and higher order in the species. The theory of evolution, therefore, must be suspended in an evidential vacuum, with no visible means of support. It flatly contradicts the universal Second Law of Thermodynamics and fails completely to meet the necessary criteria for even those superficial exceptions to the Law which occur in living organisms. Evolutionists walk by faith, not by sight!

Trying to Save Evolution from Entropy

Some evolutionists are beginning vaguely to see the devastating effects that the entropy concept can have on their theory. Certain desperate attempts have been made to salvage it, and these ideas should be mentioned briefly.

One of these efforts is to insist that, at the level of elementary particles, the Laws of Thermodynamics do not apply. Particles behave randomly, and it is thus impossible to predict their behavior. That being the case, if enough time is available, this random interplay may eventually produce a higher order strictly by chance. Anything can happen if enough time is available.

This is not true, however. Even if such random fluctuations in a chaotic assemblage of particles accidentally increased their order momentarily, further random movements would quickly destroy it. Disorder tends to increase as time goes on, by the Second Law, at least as far as an *assemblage* of particles is concerned. Nobody really knows what may or may not be true with respect to some individual fundamental particle, since our instruments are incapable of accurately resolving this question, but many physicists (Albert Einstein included) insisted that each particle also would be found to conform to the Two Laws if it could actually be traced.

In any case, when we deal with the broad subject of evolution, we are concerned with finite aggregations of particles, atoms, molecules, cells, etc., not with an isolated infinitesimal particle. All experience, as well as the equations of statistical thermodynamics, confirm that, on this scale, the entropy principle does apply in all real processes.

Another remarkable attempt to skirt the Second Law has been called *chemical predestination.* Here the attempt is made to picture the evolutionary process as inevitable, using the arguments of determinism. This means that, in the initial random concourse of atoms, properties were resident in the basic elements which somehow impelled them chemically to those unions and reactions which culminated in evolution.

This amounts to saying that there actually was a coded plan which directed the evolutionary process and that this plan was built into the structure of the elements themselves. Apart from the singular lack of evidence for this remarkable theory, it still does not deal with the real problem. Conversion of randomness to order requires "information" (or energy), a code, and a *previously-organized* mechanism for transmuting the information into the desired order. Ultimately this requires an intelligence to encode the information and devise the mechanism. Without such a primeval Intelligence, practically omniscient in understanding and foresight, the idea of chemical predestination is impossible.

This consideration brings us to a third type of attempt to avoid the anti-evolutionary implications of entropy, one that we might call *theistic cybernetics*. Here the necessity for a planning intelligence is admitted, with a divine force of some kind originating and maintaining the evolutionary process. Thus, by this provision, a continuous external supply of ordering energy is available to shore up the evolutionary process whenever entropy is about to take over.

This constitutes one of those "mystic" theories against which traditional evolutionists are perpetually polemicizing.

"The concept of teleology is in general disrepute in modern science. More frequently than not it is considered to be a mark of superstition, or at least a vestige of the nonempirical, a prioristic approach to natural phenomena characteristic of the pre-scientific era. The main reason for this discredit is that the notion of teleology is equated with the belief that future events — the goals or end-products of processes — are active agents in their own realization. In evolutionary biology, teleological explanations are understood to imply the belief that there is a planning agent external to the world, or a force imminent to the organisms, directing the evolutionary process toward the production of specified kinds of organisms."[1]

The author of the above quotation, of course, categorically rejects and deplores this idea of externally-planned evolution, as do practically all leading evolutionists today. However, the believer in a divine Intelligence behind the universe has even more reason to reject it than do these men. An intelligent Being, of such supreme power and intelligence, as to be able to pre-program and energize the development of the wonderful complexity of the organic world from the random particles of the primeval chaos, must be indeed a Being of vast wisdom and capability, for all practical purposes one who is both omniscient and omnipotent.

But then, how could we reconcile the great intelligence of such a Planner with the incredibly wasteful and inefficient process which He planned? Innumerable extinctions, misfits, blind alleys

---

[1] Francisco J. Ayala: "Biology as an Autonomous Science," *American Scientist,* Vol. 56, Autumn 1968, p. 213.

and purposeless random variation are evident in the fossil record of the supposed evolution of life, on a gigantic scale. It would seem an insult to the divine Intelligence to accuse Him of such stupidity and cruelty. No wonder Ayala says:

"The evidence of the fossil record is against any directing force, external or imminent, leading the evolutionary process toward specified goals." [1]

The theory of evolution thus leads its advocate into an impossible quandary. The process requires an intricately-complex coded plan and an efficiently-powerful energy conversion mechanism for its successful accomplishment. This in turn requires essentially an omniscient and omnipotent Planner and Provider to direct and empower the process. But surely such a Being would be too intelligent to plan it the way it has worked out and powerful enough to accomplish it in a better way, and indeed too loving and kind to allow such a spectacle of suffering and death at all! The only escape from this dilemma is finally to realize that the whole evolutionary concept can be nothing but a great delusion.

The Second Law of Thermodynamics continues to stand as an unbreached barrier against any theory of evolution. In the next several sections, its witness against each of the major stages of so-called evolution will be further detailed.

Origin of Matter and the Universe

The chemical elements constitute the atomic building blocks of matter, and evolutionists commonly attribute their formation to some catastrophic process of nucleogenesis in a primeval fireball. This concept actually is non-uniformitarian, since nothing of the sort is happening now (except for the possible build-up of hydrogen into helium in thermo-nuclear fusion processes in the stars).

"We are told that matter is being continually created, but in such a way that the process is imperceptible — that is, the statement cannot be disproved. When we ask why we should believe this, the answer is that 'the perfect cosmological principle' requires it. And when we ask why we should accept this

---

[1] Francisco J. Ayala: "Teleological Explanations in Evolution," *Philosophy of Science,* Vol. 37, March 1970, p. 11.

principle, the answer is that the principle must be true because it seems fitting to the people who assert it. With all respect, I find this inadequate."[1]

The natural tendency is for matter to disintegrate, not synthesize itself. Nuclear fission is more easily accomplished than nuclear fusion. Heavy elements undergo a natural process of radioactive decay, but not a process of naturalistic synthesis from lighter elements.

Furthermore, there is no way of accounting for the production of the primeval furnace, in which the elements could evolve, without also supposing a previous system in which entropy could be inoperative until completion of the nucleosynthesis process. All of this, of course, is speculation, completely incapable of empirical investigation.

Isaac Asimov has tried to imagine possible ways in which the entropy principle might be harmonized with the origin of the present universe. He suggests three possibilities:

"(1) . . . Somewhere, there may be changes under unusual conditions that we can't as yet study . . . in the direction of decreasing entropy.

(2) . . . It may be that through sheer random movement, a certain amount of energy concentration is piled into part of the universe. By random motion, a certain amount of order is produced once more.

(3) . . . The universe may be running down as it expands and then winding up again as it contracts, and it may be doing this over and over through all eternity."[2]

Well, *maybe!* By imagining conditions that do not prevail in our present space or time, conditions that no one has either observed or demonstrated to be possible, one could presumably "explain" anything he may fancy. But if we are limiting our discussion to *science*, the Second Law as formulated empirically absolutely precludes the formation of matter or of the universe

---

1 Herbert Dingle: "Science and Modern Cosmology," *Science,* Vol. 120, October 1, 1954, p. 515.

2 Isaac Asimov: "Can Decreasing Entropy Exist in the Universe?" *Science Digest,* May 1973, pp. 76-77.

by any *natural* process. Therefore, they must have been formed by a *supernatural* process — namely, creation!

Finally, it is significant that matter, in its present state, is always subject to defects in its structure, which make it susceptible to disintegration. The formation of crystals is sometimes cited as an exception to entropy, but it is of course based on the information already encoded within the molecular structure of the substance. Furthermore, the crystal structure is always defective in some measure, so that all materials are imperfect in some degree. In fact, these defects are themselves expressed in entropic terms.

The origin of stars is usually assumed to take place by some kind of condensation process, whereby uniformly dispersed gas cools and condenses and finally coalesces by gravity into great aggregations of material. Any such process, however, obviously demands a great decrease of entropy. By the Second Law, however, a decrease of entropy in one part of the universe requires a still greater increase of entropy somewhere else. When we are talking about the origin of stars, there *is* nowhere else! The whole idea is impossible, especially since the particles initially must be too far apart to be subject to intergravitational motions.

As far as present-day stars are concerned — and these are all science really *knows* about — they are all in a process of decay. They consume hydrogen, radiate energy to be dissipated in space, lose material in eruptions and flares. None are demonstrably in a state of growing order or complexity.

There are many theories on the origin of the solar system, but none is generally accepted and all have serious fallacies. It is significant also that the solar system is the only *known* planetary system in the whole universe.

Planets are believed to have evolved by gravitational accumulation of dust cloud particles. Again, however, this represents an increase of order, and there is no external source from which to draw the required negentropy. Before the particles are close enough to coalesce by gravity, there is no way to get them started together.

Certain other special features of the solar system such as

comets and meteorites, are of course in an obvious process of rapid decay.

Origin of Life

If entropy is a barrier to the evolution of elements and planets, it makes the origin of life from non-living chemicals an utter impossibility!

"But the most sweeping evolutionary questions at the level of biochemical genetics are still unanswered. How the genetic code first appeared and then evolved and, earlier even than that, how life itself originated on earth remain for the future to resolve, though dim and narrow pencils of illumination already play over them. The fact that in all organisms living today the processes both of replication of the DNA and of the effective translation of its code require highly precise enzymes and that, at the same time, the molecular structures of those same enzymes are precisely specified by the DNA itself, poses a remarkable evolutionary mystery."[1]

The DNA molecule and the enzymes are both extremely complex systems, and each is necessary for the other. There is no known way they could both have evolved from simpler chemicals. Dr. Haskins, President of the Carnegie Institute of Washington, continues with the following remarkable comment on this thermodynamic mystery:

"Did the code and the means of translating it appear simultaneously in evolution? It seems almost incredible that any such coincidence could have occurred, given the extraordinary complexities of both sides and the requirement that they be coordinated accurately for survival. By a pre-Darwinian (or a skeptic of evolution after Darwin) this puzzle would surely have been interpreted as the most powerful sort of evidence for special creation."[2]

Creation would indeed seem to be a logical explanation for the marvelous complexity of the simplest forms of living matter. Nothing less than the Creator could supply the astronomic gain in order represented by the simplest protein molecule.

---

[1] Caryl P. Haskins: "Advances and Challenges in Science in 1970" *American Scientist*, Vol. 59, May-June 1971, p. 298.

[2] *Ibid.*

"Directions for the reproduction of plans, for energy and the extraction of parts from the current environment, for the growth sequence, and for the effector mechanism translating instructions into growth — all had to be simultaneously present at that moment. This combination of events has seemed an incredibly unlikely happenstance, and has often been ascribed to divine intervention."[1]

In spite of all the imaginative experimentation in this field, there is still no known way in which the entropy barrier can be overcome to permit the development of living matter from non-living.

Origin of the Kinds

Not only must the evolutionary process pierce the entropy barrier to produce life, but it must then continue to pull itself onward and upward to higher and higher levels of order and complexity, all the while under the opposing pressures of the Second Law which would continually shove it back down again. To accomplish such a miracle, there surely must be at hand a tremendously powerful "metabolic motor" to force evolution to continue its upward progress. Some marvelous energizing pressure must be perpetually acting on the genetic system and its DNA molecules, imparting new and exciting "information" to them to help them evolve.

And what is this remarkable urge? This remarkable urge is the process of *random mutation*!

The very concept of a *random* process bringing order out of disorder (or transforming lower order into higher order) is itself remarkable, and is unique to evolutionary philosophy. The rules of science, mathematics, statistics, and logic work well in most fields, but for evolution we must appeal to higher insights which transcend mundane empiricism!

That mutations are indeed random changes in the genetic order is indicated by all geneticists, of which the following statement is typical:

---

[1] Homer Jacobson: "Information, Reproduction and the Origin of Life," *American Scientist,* January 1955, p. 121.

"The process of mutation ultimately furnishes the materials for adaptation to changing environments. . . . High energy radiations, such as X-rays, increase the rate of mutations. Mutations induced by radiations are random in the sense that they arise independently of their effects on the fitness of the individuals which carry them. Randomly induced mutations are usually deleterious. In a precisely organized and coupled system like the genome of an organism, a random change will most frequently decrease, rather than increase, the orderliness or useful information of the system."[1]

Mutations are sometimes called "mistakes" in the transformation of the genetic "information" encoded in the DNA structure, as it passes from parent to progeny. These mistakes are random and unpredictable as to the specific effects they will cause, both "natural" mutations and mutations artificially induced by radiations or other "mutagens."

"The artificially induced mutations are in principle of the same kind as the spontaneous ones. Thus the artificial induction of mutations can be regarded as an amplification of the natural phenomenon. It should be noted that a mutagen does not produce a specific mutation. . . . It is not possible to use mutagens to produce certain specific mutations and not other ones."[2]

It seems very strange that evolutionists should have to propose as the mechanism for introducing higher order into the organic world a process which by its very nature tends to produce disorder. Never was it ever heard, in all the world of experience, that any other kind of random process, acting on an orderly system, could elevate the order of that system!

It is no wonder that practically all mutations are deleterious in the normal struggle for existence in the natural environment. Julian Huxley has said:

"A proportion of favorable mutations of one in a thousand does not sound much, but is probably generous, since so many

1 Francisco J. Ayala: "Genotype, Environment and Population Numbers," *Science,* Vol. 162, December 27, 1968, p. 1456.
2 Bjorn Sigurbjornsson: "Induced Mutations in Plants," *Scientific American,* Vol. 224, January 1971, p. 90.

mutations are lethal, preventing the organism living at all, and the great majority of the rest throw the machinery slightly out of gear. And a total of a million mutational steps (that is, to develop a new and higher kind of organism — *author*) sound a great deal but is probably an understatement."[1]

The probability against this happening by chance is one out of a thousand raised to the millionth power. Then, Sir Julian concludes from this:

"No one would bet on anything so improbable happening and yet it has happened!"[2]

Mutations are rare and almost always harmful, and yet great numbers of favorable mutations must accumulate to produce a new kind. Thus mutation by itself is clearly incapable of overcoming entropy — in fact it is a prime example of the entropy principle in operation! How then does it produce evolution? The magic word is *natural selection*!

"Mutation is, then, the ultimate source of evolution, but there is more to evolution than mutation. . . . Mutation alone, uncontrolled by natural selection, would result in the breakdown and eventual extinction of life, not in adaptive or progressive evolution."[3]

Modern evolutionists believe that natural selection operates as sort of a sieve, which retains the very rare favorable mutations and allows the much more common unfavorable mutations to die out. Or, better, it serves as a "ratchet," which catches each successive good mutation as it passes and accumulates it on top of any previous good mutations. Some have even called it a "Maxwell demon," which is capable of setting aside the Second Law whenever it becomes convenient to do so.

Natural selection must indeed be a wonderful mechanism, turning an impossibility into a certainty and converting a universal law of decay into a universal law of growth. It is like nothing else in all the world. It can produce nothing at all by itself, of course,

---

[1] Julian Huxley: *Evolution in Action* (New York, Harper Bros. Publ., 1953), p. 41.

[2] *Ibid*, p. 42.

[3] Theodosius Dobzhansky: *Genetics of the Evolutionary Process* (New York, Columbia University Press, 1970) p. 65.

but must merely wait patiently for the mutation process to activate it at rare intervals. It has no program to guide it, no energy to sustain it, and yet, as we have seen, all other known growth processes require both a pre-designed program and a complex energizer.

Actually, natural selection is a conservation mechanism tending to conserve the kinds already established and to eliminate the normally deleterious results of the mutation process, (which is of course a perfect example of a decay process). Natural selection and mutation illustrate vividly the basic laws of conservation and decay. It is wishful thinking, at best, to assume they provide the means of evolutionary progress.

The Ubiquitous Entropy Principle

Other illustrations of the decay principle abound throughout the world of nature and life. A few of these are mentioned briefly below.

(1) *Frictional resistance to motion.* Everything happening in our space-time universe involves motion, and every motion requires a force to initiate it against inertia and sustain it against the retarding force of friction. There is always some friction resisting the movement, and some of the energy maintaining it must be used to overcome the friction, being converted thereby into non-recoverable heat energy. Thus all processes tend eventually to slow down and cease. Perpetual motion is impossible and no process can be operated at 100% efficiency.

(2) *Aging and death of individuals.* Individual organisms of all kinds eventually die, no matter how much energy may be available to them in the environment. Body cells deteriorate, complex protein molecules break down into simpler compounds, and eventually some vital organ or process ceases to function and death ensues. The causes of the aging process are not yet well understood, but there seems no basic reason why it could not be slowed down considerably with resulting increased longevity. The Second Law does not specify the rate, but only the fact, of decay.

(3) *Decay and extinction of species.* Just as individuals die,

so do species. The real testimony of the fossil record is not the slow development of species, but rather the sudden extinction of species. Similarly, there may be such things as *vestigial* organs in both extinct and living kinds of animals, but no one is able to point to any *nascent* (that is, beginning to evolve into something useful) organs.

(4) *Wear*. Another result of friction is that of wear. Not only does friction convert useful energy into useless heat, but also useful material into dust.

"The inexorable loss of material from surfaces in sliding contact eventually destroys the usefulness of most things."[1]

(5) *Disease*. Both plants and animals are subject to still another kind of deterioration, namely, disease.

"Long before man began to evolve about one million years ago, the earth was inhabited at least half a billion years ago by numerous other animals: the fishes, the reptiles, and eventually the mammals, many of which are now extinct. These prehistoric animals, scientists say, often were afflicted with disease."[2]

In other words, and translating the above evolutionary terminology into a creationist context, this type of decay has been present since nearly the beginning of life on the earth.

(6) *Decay of environment*. Not only do both individual organisms and even entire species decay and die, but also the very environment of life tends to decay. People have become especially aware of this factor in connection with pollutional problems in recent years, but actually the environment is continually decaying even apart from man-made pollution. Soil erosion, for example, continually contributes to stream and ocean solution chemicals, quite independently of artificial contamination thereof. Climatic factors are constantly changing also.

(7) *Decay of nations, cultures, and languages*. Although it

---

1 Ernest Rabinowicz: "Wear," *Scientific American*, Vol. 206, February 1962, p. 127.

2 James A. Tobey: "Disease is Older than Man," *Science Digest*, April 1958, p. 53.

might be difficult to demonstrate a formal connection with thermodynamics, there is a marked analogy between the birth, growth, and death of an individual, and the rise and fall of nations and cultures. Languages also typically become decadent and simple in their old age.

(8) *Breakdown of morals and religion.* Similarly, societies commonly experience a period of strong religious faith and strict moral standards in the early days of their development, only to be gradually affected by an apparently inevitable decline in faith and morals as they become older.

(9) *Personal disintegration.* Finally, a universal fact of experience is that one's moral behavior is under constant pressure to drop down to lower levels. If one simply "lets himself go," he inevitably goes down, morally and spiritually. He does not automatically get better. Even such a godly man as the Apostle Paul had to say: "I find then a law, that, when I would do good, evil is present with me. For I delight in the law of God after the inward man. But I see another law in my members, warring against the law of my mind, and bringing me into captivity to the law of sin which is in my members. O wretched man that I am! Who shall deliver me from the body of this death?" But he also gives the answer: "I thank God through Jesus Christ our Lord." (Romans 7:21-25).

We conclude this chapter by contrasting the entropy principle with the Creator to whom it points. The Laws of Thermodynamics, as already discussed, require the original creation of the universe by a Creator transcendent to it. The Creator, therefore, who established the Two Laws, is not Himself subject to them. This is emphasized in a remarkable passage of Scripture, Isaiah 40:26-31.

"Lift up your eyes on high, and behold who hath created these things, that bringeth out their host by number (creation of ORDER): He calleth them all by names by the greatness of His might (creation of INFORMATION), for that He is strong in power, not one faileth (creation of POWER)."

He conserves them all (First Law), and even though they may "faint" (Second Law), the Creator Himself is unchanged.

"Hast thou not known? Hast thou not heard, that the everlasting God, the Lord, the Creator of the ends of the earth, fainteth not (no loss of Order), neither is weary (no loss of Energy)? There is no searching of His understanding" (no loss of Information).

Even though now the "whole creation groaneth and travaileth in pain" under the "bondage of decay" to which the "creation was made subject" (Romans 8:22, 21, 20), the Creator will not be defeated in His purpose in creation and will someday "make all things new" (Revelation 21:5). This renovation awaits the Second Coming of Christ who, at His First Coming, paid the price, "the blood of His cross, by Him to reconcile all things unto Himself" (Colossians 1:20).

Individually, even now, the reconciliation and restoration is available to all who place their faith in Him.

"But they that wait upon the Lord shall renew their strength; (literally, 'exchange their strength' — talk about an energy conversion and conservation — our weakness for His eternal power!); they shall mount up with wings as eagles; they shall run, and not be weary; and they shall walk and not faint" (Isaiah 40:31).

The evolutionary delusion has not only had a profound influence in shaping human thought in every field in the past, but is also now being applied by its devotees to the great problems confronting the world today, with ominously turbulent results.

Norris Geyser Basin, Yellowstone National Park, Wyoming

Photo credit: Col. M. G. McBee

CHAPTER VI

BOIL AND
BUBBLE

In previous chapters we have traced the long history of evolutionary thought, and its ubiquitous influence in modern society. We have also noted its complete failure in light of the facts of science.

In this chapter, we wish to note how evolutionary thinking has been applied in relation to various important specific problems in the modern world. A number of different topics will be discussed, the point of commonality being their current interest and future importance. Each of these problems has become critical in one way or another, due primarily to the confusing influence of evolutionist theory and methodology in their study and application. Creationist thought would be found much more salutary in each case.

Evolution and the Population Problem

Few issues today are more emotionally charged than that of population control. Sociological alarmists insist that the growth of human populations must be stopped by whatever means are available. Not only the usual contraceptive methods, but even such anti-Scriptural practices as abortion and homosexuality have been promoted as desirable to help attain the goal of zero population growth.

The intellectual and educational establishments today assume it as self-evident that population growth should be halted. Famed anthropologist Margaret Mead, in the lead editorial in a recent issue of *Science*, says:

"The United Nations Population Conference, which concluded on 31 August in Bucharest, passed by acclamation a World Plan of Action that dramatized the growing global concern for the planet's plight. . . . At Bucharest it was affirmed that continuing, unrestricted worldwide population growth can negate any socio-economic gains and fatally imperil the environment. . . . Those governments for which excessive population growth is detrimental to their national purpose are given a target date of 1985 to provide information and methods for implementing these goals." [1]

The cause for such concern is the current high rate of world population increase and the fear that this presages the imminent depletion of the world's capacity to sustain its inhabitants. Two of the leading specialists in this field are Donald Freedman, Professor of Sociology at the University of Michigan, and Bernard Berelson, President of the Population Council. These authorities define the problem as follows:

"In the 1970's the rate of increase has slightly exceeded 2 per cent per year. That means a doubling time of less than 35 years, and the number currently being doubled is a very large one. Projection of such growth for very long into the future produces a world population larger than the most optimistic estimates of the planet's carrying capacity." [2]

So urgent do the experts consider this problem to be that the United Nations Organization actually proclaimed 1974 to be "World Population Year." It can be shown, in fact, that if the

---

[1] Margaret Mead: "World Population: World Responsibility," *Science,* Vol. 185, September 27, 1974, p. 1113.

[2] Donald Freedman and Bernard Berelson: "The Human Population," *Scientific American,* Vol. 231, September 1974, p. 31.

population continued to increase at the rate of 2% per year, in less than 700 years there would be one person for every square foot of the earth's surface. Obviously, the present growth rate cannot continue indefinitely.

Nevertheless, many creationists find such arguments unconvincing. Since the evidence for a purposeful Creator of the world and mankind is exceedingly strong, the creationist is confident that the world God made for man is large enough and productive enough to accomplish His purpose. That purpose will surely have been consummated before the population exceeds its divinely-intended maximum.

According to the Biblical record of creation, immediately after the first man and woman were created, God instructed them as follows:

"Be fruitful and multiply, and replenish (literally, 'fill') the earth, and subdue it." (Genesis 1:28). Essentially the same commandment was given to the handful of survivors of the great Flood (Genesis 9:1). Since man has not yet come anywhere near to *filling* the earth (the total population currently averages less than one person for every 400,000 square feet of land area), it seems unlikely that the earth has yet reached its optimal population, as far as the purposes of the Creator are concerned. The divine command no doubt at least envisioned colonizing all parts of the earth and occupying each part to its potential maximum productive capacity.

Throughout the Scriptures, a large family is considered to be a blessing from the Lord (note Psalm 127:3-5; 128:1-6, etc.), not a problem to society, assuming, of course, that these children are going to be brought up "in the nurture and admonition of the Lord" (Ephesians 6:3).

The historic fact of creation is prophetic of the future fact of consummation. Many current trends seem to have been predicted in the Bible and, therefore, suggest that the return of Christ and the end of the age may be near at hand. It is, therefore, at least a possibility that the Creator's work of consummation may solve the population problem long before it becomes critical.

Even apart from Biblical revelation, however, there is no good reason for alarm over population. The earth is quite able to support a much larger population than it now possesses. Even with the present status of technology (available water for irrigation, potentially arable land, modern methods of soil treatment and improved crop yields, etc.), authorities estimate that the earth's reasonable "carrying capacity" is about 50 billion people.[1] Future advances in technology (solar energy, saline conversion, etc.), may well increase this still more.

Thus, even at the present annual increase of 2%, it will still be 135 years before this maximum population will be reached. However, in order for such a population to be achieved, modern technological knowledge will have to be employed worldwide, in the underdeveloped countries as well as in the developed nations. In turn, experience in the latter shows that population growth rates tend to drop off as a society's technology increases. Revelle comments on this as follows:

"Here we are faced with a paradox: attainment of the earth's maximum carrying capacity for human beings would require a high level of agricultural technology, which in turn calls for a high level of social and economic development. Such developments, however, would be likely to lead to a cessation of population growth long before the maximum carrying capacity is reached."[2]

It is interesting that, for the most part, those intellectuals who are most vocal in support of population limitation (Margaret Mead, for example), are also strong believers in human evolution. This is probably because of their refusal to recognize divine purpose in the world. If there was no creation and therefore no purpose or goal in creation, as leading evolutionists believe, then neither is there any reason to believe the Creator will accomplish His purpose at the end of history. Just as man's past evolution was dependent solely on random natural processes, so must his future be controlled by naturalism, the only difference being that man now knows how to control those processes — or so he hopes.

One of the strange aberrations of the modern drive for

---

1 Roger Revelle: "Food and Population," *Scientific American,* Vol. 231, September 1974, p. 168. Revelle is Director of the Center for Population Studies at Harvard.

2 *Ibid.,* p. 169.

ecological and population controls is the notion that the "environmental crisis" is an outgrowth of the Biblical teaching that man should multiply numerically and subdue the earth. Professor Lynn White of U.C.L.A. first popularized the notion that this Genesis mandate has served as man's justification for the exploitation of the earth's resources.[1] Professor Richard Means and others have even proposed that we should all revert to belief in a pantheistic polytheism in order to have a proper regard for all aspects of the world and its living things as they have evolved. [2]

This idea is a prime example of evolutionistic confusion of thinking. Christian scholars have never used Genesis 1:28 in support of the careless use and waste of any of the earth's resources. To the contrary, since everything is presented in Scripture as the product of God's creative design and purpose, Biblical creationist Christians regard themselves and man in general as stewards of the whole creation, accountable to the Creator for its proper development and use.

On the other hand, it is very significant that all of the earth's serious environmental problems, even its population crisis, have developed during that one century when the evolutionary philosophy had effectively replaced creationism in the thinking of the world's leaders in education, science, and industry. The earth has been exploited not because of any divine mandate, but because of social Darwinism, economic and military imperialism, secular materialism, anarchistic individualism, and other such applications of the "struggle and survival" rationale of modern evolutionism.

As far as reverting to pantheism is concerned, this is simply another variant of evolutionism and will inevitably lead to similar results. The most pantheistic of nations (e.g., India with its Hinduism, China with its Buddhism and Confucianism, etc.) are precisely those nations in which the population/resource ratios have been most severe. It has not been the Judaeo-Christian nations in which population has become a problem, but those with

[1] Lynn White: "The Historical Roots of our Ecological Crisis," *Science,* Vol. 155, March 10, 1967, pp. 1203-1207.
[2] Richard L. Means: "Why Worry about Nature," *Saturday Review,* December 2, 1967.

religions of pantheism. How then can pantheism solve the very problems it nurtures?

But there is an even greater inconsistency in evolutionary thinking relative to population. The same population statistics which supposedly presage a serious population problem in the future also indicate a very recent origin of man in the past.

To illustrate the problem, assume that the human population increases geometrically (as believed by Malthus, whose writings were of profound influence on the theories of Charles Darwin). That is, the increase each year is equal to a constant proportion of the population the previous year. This relationship can be expressed algebraically as follows:

$$P_n = P(1+r)^n \qquad (1)$$

in which P is the population at any certain time, r is the proportionate annual increase in population, and P_n is the population n years later. For example, if the present population is 3.5 billion and the planet's permissible population is 50 billion, the number of years before this number will be reached at the present 2% annual increase can be calculated as follows:

$$50 \times 10^9 = 3.5 \times 10^9 \, (1.02)^n$$

from which

$$\log \frac{50}{3.5} = n \log 1.02$$

and

$$n = \frac{1.156}{0.0086} = 135 \text{ years.}$$

We have already discussed this result, however. Looking toward the past, instead of the future, equation (1) will also indicate how long it would take to produce the present population at 2% growth per year, starting with two people. Thus:

$$3.5 \times 10^9 = 2(1.02)^n$$

from which

$$n = \frac{9 + \log \left(\frac{3.5}{2} \right)}{\log 1.02} = 1075 \text{ years}$$

That is, an initial population of only two people, increasing at 2% per year, would become 3.5 billion people in only 1075 years. Since written records go back over 4,000 years, it is obvious that the average growth rate throughout past history has been considerably less than the present rate.

As a matter of interest, we can also use equation (1) to determine what the average growth rate would have to be to generate the present population in 4,000 years. Thus:

$$3.5 \times 10^9 = 2(1+r)^{4000}$$

from which $\qquad r = (1.75 \times 10^9)^{\frac{1}{4000}} -1 = \frac{1}{2}\%$

Thus, an average population growth rate of only ½% would generate the present world population in only 4000 years. This is only *one-fourth* of the present rate of growth.

Now, although it is obvious that the present rate of growth (2%) could not have prevailed for very long in the past, it does seem very unlikely that the growth rate could have averaged significantly less than ½%. Very little is known about the world population in earlier times, but everything that *is* known indicates the population has steadily increased throughout recorded history.

Dr. Ansley J. Coale, Director of the Office of Population Research at Princeton University, has discussed the paucity of such data in an important recent study.

"Any numerical description of the development of the human population cannot avoid conjecture, simply because there has never been a census of all the people of the world . . . . The earliest date for which the global population can be calculated with an uncertainty of only, say 20 per cent is the middle of the 18th century. The next earlier time for which useful data are available is the beginning of the Christian era, when Rome collected information bearing on the number of people in various parts of the empire."[1]

---

1 A. J. Coale: "The History of the Human Population," *Scientific American,* Vol. 231, September 1974, p. 41.

The usually-accepted estimates of world population for these two dates are, respectively, about 200 million in A.D. 1 and about one billion in 1800 A.D. The first, however, may be vastly in error, since no one really knows the population in most parts of the world at that early date.

For earlier periods than A.D. 1, absolutely *nothing* is *known* concerning world populations. It should be emphatically stressed that *all* estimates of earlier populations except that recorded in the Bible (namely, that immediately after the great Flood, the world population consisted of eight people) are based solely on evolutionary concepts of human technological development.

"For still earlier periods (than A.D. 1) the population must be estimated indirectly from calculations of the number of people who could subsist under the social and technological institutions presumed to prevail at the time. Anthropologists and historians have estimated, for example, that before the introduction of agriculture the world would have supported a hunting and gathering culture of between five and ten million people."[1]

Such guesses are useless, however, because they are based on a discredited model, that of human evolution. The creation-cataclysm model of earth history fits all the known facts of man's history much better than the evolution model does,[2] and it recognizes that man's agriculture and other basic technologies are essentially as old as man himself.

In 1650 the world population has been estimated with perhaps reasonable accuracy to have been 600 million. In 150 years this had grown to approximately one billion. The average rate of increase for this period, therefore, is:

$$r = (\frac{10}{6})^{\frac{1}{150}} - 1 = 1/3\%$$

Since this period from 1650 and 1800 antedated the great advances in medicine and technology which have stimulated the more rapid population growth of the 19th and 20th centuries, and

---

[1] *Ibid.*, p. 41.

[2] See *Scientific Creationism* (Ed. by Henry M. Morris, San Diego, Creation-Life Publishers, 1974), pp. 171-201.

also since this is the earliest period of time for which population data are at all reliable, it seems reasonable that this figure of 1/3%, rather than the ½% previously calculated, should be used as the norm for population growth throughout most of past history.

In that case, the length of time required for the population to grow from 2 people to one billion people, at 1/3% increase per year is:

$$n = \frac{\log\left(\frac{10^9}{2}\right)}{\log(1.00333)} = 6100 \text{ years}$$

To this should be added the 175 years since 1800. Thus, the most probable date of man's origin, based on the known data from population statistics, is about 6,300 years ago.

This figure, of course, is vastly smaller than the usually assumed million-year history of man. Nevertheless it correlates well not only with Biblical chronology but also with other ancient written records as well as with even the usual evolutionary dates for the origin of agriculture, animal husbandry, urbanization, metallurgy and other attributes of human civilization.

By arbitrary juggling of population models, of course, the evolutionist can manage to come out with any predetermined date he may choose. People should realize, however, that this does require an arbitrary juggling of figures, based solely on the assumptions of human evolution. The actual data of population statistics, interpreted and applied in the most conservative and most probable manner, point to an origin of the human population only several thousands of years ago. The present population could very easily have been attained in only about 6000 years or so, even if the average population growth rate throughout most of history were only one-sixth as much as it is at present. The burden of proof is altogether on evolutionists if they wish to promote some other population model.

The Biblical model for population growth starts with eight people (Noah, his three sons, and their wives) right after the great Flood. The date of the Flood is not certain; the Ussher chronology dates it about 2350 B.C., but possible gaps in the genealogies of

Genesis 11 may justify a date as far back as, say, about 6000 B.C., with the probabilities favoring the lower limit rather than the upper limit.

Even using the short Ussher chronology, it is quite reasonable, as we have seen, for the population to have grown from 8 people to 3.5 billion people in 4350 years. This growth represents an average annual increase of only 0.44%, or an average doubling time of 152 years. Such figures are quite consistent with all known data of population statistics, especially in light of the fact that the human death rates were very low for many centuries after the Flood, and family sizes quite large. Thus, in all likelihood, the population growth was very substantial in the early centuries, at least as great as it has become in the present century. In turn, this means that the rate may have been much less than 0.44% during the long period in between.

In any case, the conclusion is well justified that the Biblical chronology, even in its most conservative form, fits well into all the known facts of population growth, much more so than does the evolutionary chronology of human history.

Evolution, Energy, and Ecology

One of man's most vexing problems today is the conflict between energy and ecology, between conservation of jobs and conservation of nature. The need for expanded energy sources and more goods and services for mankind seems completely at cross purposes with the maintenance of an unpolluted environment.

Is there a way out of this dilemma? If not, what is wrong with a world that forces us into such a situation?

As with all great issues, the way in which a person views a problem and the course of action he follows in handling it depend fundamentally upon his basic philosophy of life. The ecological crisis, in particular, points up the evolution-creation conflict in a surprising light.

In the first place, the ecological relationships between groups of organisms and their respective environments, or "ecological niches" would seem to indicate intelligent forethought and planning, not random struggle between populations. The innumerable and remarkable "adaptations" of this sort are very difficult to "explain" in evolutionist terms. Two of the world's leading biological students of ecology have pointed this out:

"Some biologists claim that an understanding of the evolutionary history of organisms is prerequisite to any comprehension of ecology. We believe that this notion is having the effect of sheltering large areas of population biology from the benefits of rigorous thought. . . . Indeed, since the level of speculation rather than investigation is inevitably high in phylogenetic studies of any kind, a preoccupation with the largely unknown past can be shown to be a positive hindrance to progress."[1]

Thus, a supposed understanding of evolution is of no value, and its theories even harmful, to a true understanding of modern ecological relationships. Evolutionary ideas concerning these matters, furthermore, are based on theory, not on fact.

"Indeed, we know nothing whatever of the antecedents of most species for thousands of years. Perhaps these dismal facts account for some of the strangely unsatisfying 'explanations' of the evolutionary ecologists."[2]

Since evolution provides no satisfactory explanation of present ecological relationships, it obviously can provide no guidance for future policies on environmental problems. Creationism provides better solutions as to both past developments and future guidelines.

However, evolutionists have in recent years been propagating the absurd notion that man's exploitation of the world's resources has been based on the supposed Biblical teaching that these had been made strictly for this purpose. The Bible, however, teaches no such thing. If there is to be any placing of blame for the problem of pollution and related ills, it should be assigned to the philosophy of evolution, where it really belongs. Furthermore, effective remedies for such problems can be found only in the context of a sound creationist philosophy.

[1] Paul R. Ehrlich and L. C. Birch: "Evolutionary History and Population Biology," *Nature*, Vol. 214, April 22, 1967, p. 349.
[2] *Ibid*, p. 351.

The essence of evolution, of course, is randomness. The evolutionary process supposedly began with random particles and has continued by random aggregations of matter and then random mutations of genes. The fossil record, as interpreted by evolutionists, is said by them to indicate aeons of purposeless evolutionary meanderings, the senseless struggling and dying of untold billions of animals, extinctions of species, misfits, blind alleys. The present-day environmental-ecologic complex then is nothing more than the current stage in this unending random struggle for existence.

Those populations of organisms which have survived to this point therefore must represent the "fittest" — those that have been screened and preserved by the process of natural selection. In spite of its randomness, therefore, evolutionists believe that the net result of evolution has somehow been the development of higher and higher kinds, and finally of man himself.

This development is believed by most evolutionists to have been made possible by a peculiar combination of small populations, changing environments, and accelerated mutational pressures, a combination which supposedly enables natural selection to function in its remarkable role as "creator" of new and better kinds of populations. It would seem therefore that anything that would change the environment today (for example, by altering the chemical components of the atmosphere and hydrosphere through pollution), decrease populations (perhaps by war, famine, or pestilence), or increase mutational pressures (such as by increasing the radioactive component of the biosphere through nuclear testing), would contribute positively to further evolution and therefore should be encouraged, at least if evolutionists are correct in their understanding of evolutionary mechanisms. In other words, the very processes which modern ecologists most deplore today are those which they believe to have been the cause of the upward evolution of the biosphere in the past. The conclusion would seem to be that evolution requires pollution!

More directly to the point, however, three generations of evolutionary teaching have had the pragmatic result of inducing in man an almost universal self-centeredness. God, if He exists at all, is pushed so far back in time and so far out in space that men no longer are concerned about responsibility to Him. As far as other people are concerned, doesn't nature itself teach that we must struggle and compete for survival? Self-preservation is nature's first law. Race must compete against race, nation against nation, class against class, young against old, poor against rich, man against man.

Two thousand years of Christian teaching linger on to some extent, in modern social concerns and in the diluted esthetics and ethics of the day, but these are easily forgotten when one's self-interests are at stake. Conservationist groups may inveigh against the ecological destruction wrought by the oil and utilities industries, but they do not personally wish to give up their automobiles or electrical appliances, nor to pay the higher prices required if these commodities are to be produced without damage to the environment. Furthermore, during the past 150 years especially, the very exploitation of nature — its flora and fauna, its resources, and even its human populations — against which environmentalists are protesting, has itself been carried out in the name of science and evolutionary philosophy. Thus the modern ecologic crisis is not a product of Biblical theology at all, but rather a century of worldwide evolutionary thinking and practice. It is significant that all of these environmental problems developed almost entirely within a period when the scientific and industrial establishments were totally committed to an evolutionary philosophy!

Recognition of the world as God's direct creation, on the other hand, transforms man's outlook on nature and his attitude toward other men. The creation is God's unique handiwork and displays His character and glory (Psalm 19:1; Psalm 148; Rev. 5:13). The design and implementation of this marvelous universe and its varied inhabitants were to God a source of great delight (Genesis 1:31; Job 38:4-7; Revelation 4:11). Man was created, not to exploit God's world, but to be His steward, exercising dominion over it (Genesis 1:26-28) and "keeping" it (Genesis 2:15).

The primeval world as it came from God's hand was beautiful beyond imagination and perfect in every way as man's home. There was ample food for both man and animals (Genesis 1:29, 30) and each kind had its own ecological niche. Even when God stopped creating (Genesis 2:1-3), He provided abundantly for the maintenance of the creation (Nehemiah 9:6; Hebrews 1:3).

With man's fall and God's Curse on his dominion, this pristine perfection changed (Genesis 3:17; Romans 8:20, 22). Every process henceforth operated inefficiently, and every system tended toward disintegration. Although the earth's resources remained constant in quantity, their quality could thereafter be maintained only with great difficulty and only at the cost of drawing excess energy from some other source.

Not only was the quantity of matter and energy originally intended to be "conserved" (as expressed formally now in our scientific Law of Conservation of Energy), but presumably also the "quality" of energy. Not only was energy conserved, but entropy as well; the universe was not designed to "perish," "wax old," and "be changed" (Psalm 102:26; Psalm 148:6), but to be "stablished for ever and ever." In some unknown manner, no longer operating, the sun's energy probably was replenished cyclically, from that radiated into space after some had been used to maintain terrestrial processes. On the earth itself, none of its resources were ever to be depleted and all processes were to function at perfect efficiency. A great abundance of plant and animal life was soon produced, in response to God's commands (Genesis 1:11, 20, 24), and continued to multiply, storing energy from the sun in an enlarging biosphere. All necessary disintegrative processes (e.g., digestion, etc.) were presumably in balance with the increasing numbers of highly structured organisms. Order and entropy were thus everywhere in balance, as well as matter and energy. Everything was "very good" (Genesis 1:31).

The Bible gives little information as to such specific energy sources before the Flood, except for the sun itself. At the time of the Deluge, however, the earth's energy balance changed drastically. Its greenhouse-like environment, which had been maintained by "waters above the firmament" (Genesis 1:7), was destroyed when the great canopy of vapor condensed and deluged the entire globe. The tremendous stores of chemical energy in the biosphere of the antediluvian world were partially converted in the resulting cataclysm into great stores of coal, oil and gas, the so-called "fossil fuels." Much of the incoming solar energy thenceforth would be needed to drive the atmospheric circulations and to maintain the post-diluvian hydrologic cycle for the earth.

It is significant to realize that today's pollution problems are derived mostly from using energy stores that were produced in the Noachian Deluge! Coal is the fossil product of the terrestrial plant life, and oil largely of the marine animal life, of the rich biosphere that had been created and developed by the Creator in the beginning. These organisms were not designed to serve as fuels for man's machines, and it is not surprising that the efficiency of heat engines using them is low and the waste products are high. Furthermore, they are exhaustible and, even now, the imminent end of economical oil and gas production is a matter of great concern.

In a sense, of course, the burning of these fossil fuels is merely hastening the process of "returning to the dust," which is the present fate of all organic life, under the Curse. The waste products, both of the processes of life and of the phenomena of death, have always posed a pollution problem to the environment, but the normal cycles of nature are able to accommodate them in part, and even utilize them (e.g., in the enrichment of the soil, etc.) as long as they are sufficiently dispersed in time and space. When concentrated in abnormal numbers of either men or animals, either in living communities or massive extinctions, however, such wastes cannot be assimilated and initiate various abnormal reactions which accelerate and accentuate environmental decay.

These deleterious changes can be corrected to some extent, but only at the cost of excess energy from other sources and therefore only at great labor and expense. Nuclear energy is one possibility but this of course creates its own pollutional problems. Geothermal energy may be a partial answer, in the few regions where it is available. Hydroelectric energy has already been developed to nearly its maximum potential in many parts of the world and is seriously limited in all parts of the world. The energy in the tides and ocean waves is considerable, but its harnessing is economically feasible only in very restricted localities.

Solar energy is undoubtedly the best ultimate hope for an adequate energy supply, since the sun is the ultimate source of energy for all of earth's processes anyhow. To date, however, no economically efficient solar converters have been developed, except for special and limited applications. Since the sun was created to "give light upon the earth" (Genesis 1:17) and since "there is nothing hid from the heat thereof" (Psalm 19:6), we may well believe that it is possible to find ways to utilize solar energy to meet all of man's legitimate energy needs and to do so with a minimal amount of further damage to the environment. Cost of the needed research should not be prohibitive, at least in relation to other energy and environmental costs.

In any case, a creationist orientation can certainly contribute more effectively to the alleviation of such problems than can an evolutionary perspective. The creationist recognizes that the world is God's handiwork and that he is God's steward. The divine commission to "have dominion over" and to "subdue" the earth is not a license for despotic exploitation of its resources, but rather a call to service, encouraging him to understand its nature ("science") and then to utilize its resources ("technology") for the benefit of all men, under God.

Eventually, however, if the present world (no matter how carefully its resources were guarded) were to continue indefinitely operating under the present laws of nature, it would die. The "whole creation" is under the "bondage of decay" (Romans 8:20-22).

But this bleak prospect will never be reached. God's eternal purpose in creation cannot fail. The creation, therefore, must be somehow redeemed and saved. Although in the present order, the Curse is universal and inexorable, the One who imposed it can also remove it (Revelation 22:3).

The redemption price has in fact been paid in full (Colossians 1:20) and this "redemption of the purchased possession" (Ephesians 1:14) will be completely implemented when Christ returns. At that time, everything, including the earth and its land-water-air environment, will all be "made new" again (Revelation 21:5), and will then last forever.

In the meantime, every person who has appropriated this redemption individually through an act of faith in his Creator and Redeemer has the privilege of sharing in God's work of reconciliation, for "He hath given to us the ministry of reconciliation (II Corinthians 5:18). The work of redemption and reconciliation involves the reclamation and saving both of individual men and of man's dominion, for the eternal ages to come. We seek not only to win scientists to Christ, but even to win the sciences themselves to Christ.

"O Lord, how manifold are thy works! in wisdom hast thou made them all; the earth is full of thy riches. . . . The glory of the Lord shall endure for ever; the Lord shall rejoice in His works" (Psalm 104:24, 31).

Evolution and Modern Racism

One of the most vexing problems facing modern man is that of conflict between the various races. This problem was discussed briefly in Chapter II. At this point, we wish to consider it once more, this time as one of those modern problems which evolutionary thinking has aggravated and which could be solved by the application of true Biblical creationism.

Some people today, especially those of anti-Christian opinions, have the mistaken notion that the Bible prescribes permanent racial divisions among men and is, therefore, the cause of modern racial hatreds. As a matter of fact, the Bible says nothing whatever about race. Neither the word nor the concept of different "races" is found in the Bible at all. As far as one can learn from a study of Scripture, the writers of the Bible did not even know there were distinct races of men, in the sense of black and yellow and white races, or Caucasian and Mongol and Negroid races, or any other such divisions.

The Biblical divisions among men are those of "tongues, families, nations, and lands" (Genesis 10:5, 20, 31) rather than races. The vision of the redeemed saints in heaven (Revelation 7:9) is one of "all nations, and kindreds, and people, and tongues," but no mention is made of "races." The formation of the original divisions, after the Flood, is based on different languages (Genesis 11:6, 9), supernaturally imposed by God, but nothing is said about any other physical differences.

Some have interpreted the Noahic prophecy concerning his three sons (Genesis 9:25-27) to refer to three races — Hamitic, Semitic and Japhetic — but such a meaning is in no way evident from the words of this passage. The prophecy applies to the descendants of Noah's sons, and the various nations to be formed from them, but nothing is said about three races. Modern anthropologists and historians employ a much different terminology than this simple trifurcation for what they consider to be the various races among men.

Therefore, the origin of the concept of "race" must be sought elsewhere than in the Bible. If certain Christian writers have interpreted the Bible in a racist framework, the error is in the interpretation, not in the Bible itself. In the Bible, there is only one race — the human race! "(God) hath made of one blood all nations of men" (Acts 17:26).

In modern terminology, a race of men may involve quite a large number of individual national and language groups. It is, therefore, a much broader generic concept than any of the Biblical divisions. In the terminology of biological taxonomy, it is roughly the same as a "variety," or a "sub-species." Biologists, of course, use the term to apply to sub-species of animals, as well as men.

For example, Charles Darwin selected as the sub-title for his book *Origin of Species* the phrase "The Preservation of Favoured Races in the Struggle for Life." It is clear from the context that he had races of animals primarily in mind, but at the same time it is also clear, as we shall see, that he thought of races of men in the same way.

That this concept is still held today is evident from the following words of leading modern evolutionist George Gaylord Simpson:

"Races of men have, or perhaps one should say 'had,' exactly the same biological significance as the sub-species of other species of mammals." [1]

---

[1] George Gaylord Simpson: "The Biological Nature of Man," *Science,* Vol. 152, April 22, 1966, p. 474.

It is clear, therefore, that a race is not a Biblical category, but rather is a category of evolutionary biology. Each race is a sub-species, with a long evolutionary history of its own, in the process of evolving gradually into a distinct species.

A recent study suggests that each of the races has been around a long while.

"The simplest interpretation of this conclusion today would envision a relatively small group starting to spread not long after modern man appeared. With the spreading, groups became separated and isolated. Fifty thousand years or so is a short time in evolutionary terms, and this may help to explain why, genetically speaking, human races show relatively small differences." [1]

As applied to man, this concept, of course, suggests that each of the various races of men is very different, though still inter-fertile, from all of the others. If they continue to be segregated, each will continue to compete as best it can with the other races in the struggle for existence and finally the fittest will survive. Or else, perhaps, they will gradually become so different from each other as to assume the character of separate species altogether (just as apes and men supposedly diverged from a common ancestor early in the so-called Tertiary Period).

Most modern biologists today would express these concepts somewhat differently than as above, and they undoubtedly would disavow the racist connotations. Nevertheless, this was certainly the point of view of the 19th century evolutionists, and it is difficult to interpret modern evolutionary theory, the so-called neo-Darwinian synthesis, much differently.

The rise of modern evolutionary theory took place mostly in Europe, especially in England and Germany. Europeans, along with their American cousins, were then leading the world in industrial and military expansion, and were, therefore, inclined to think of themselves as somehow superior to the other nations of the world. This opinion was tremendously encouraged by the concurrent rise of Darwinian evolutionism and its simplistic

---

[1] L. L. Cavalli-Sforza: "The Genetics of Human Populations," *Scientific American,* Vol. 231, September 1974, p. 89.

approach to the idea of struggle between natural races, with the strongest surviving and thus contributing to the advance of evolution.

As the 19th century scientists were converted to evolution, they were thus also convinced of racism. They were certain that the white race was superior to other races, and the reason for this superiority was to be found in Darwinian theory. The white race had advanced farther up the evolutionary ladder and, therefore, was destined either to eliminate the other races in the struggle for existence or else to have to assume the "white man's burden" and to care for those inferior races that were incompetent to survive otherwise.

Charles Darwin himself, though strongly opposed to slavery on moral grounds, was convinced of white racial superiority. He wrote on one occasion as follows:

"I could show fight on natural selection having done and doing more for the progress of civilization than you seem inclined to admit. . . . The more civilized so-called Caucasian races have beaten the Turkish hollow in the struggle for existence. Looking to the world at no very distant date, what an endless number of the lower races will have been eliminated by the higher civilized races throughout the world"[1]

The man more responsible than any other for the widespread acceptance of evolution in the 19th century was Thomas Huxley. Soon after the American Civil War, in which the negro slaves were freed, he wrote as follows:

"No rational man, cognizant of the facts, believes that the average negro is the equal, still less the superior, of the white man. And if this be true, it is simply incredible that, when all his disabilities are removed, and our prognathous relative has a fair field and no favour, as well as no oppressor, he will be able to compete successfully with his bigger-brained and smaller-jawed rival, in a contest which is to be carried out by thoughts

---

[1] Charles Darwin: *Life and Letters, 1,* letter to W. Graham, July 3, 1881, p. 316; cited in *Darwin and the Darwinian Revolution,* by Gertrude Himmelfarb (London, Chatto and Windus, 1959), p. 343.

and not by bites."[1]

Racist sentiments such as these were held by all the 19th century evolutionists. A recent book[2] has documented this fact beyond any question.

In a day and age which practically worshipped at the shrine of scientific progress, as was true especially during the century from 1860 to 1960, such universal scientific racism was bound to have repercussions in the political and social realms. The seeds of evolutionary racism came to fullest fruition in the form of National Socialism in Germany. The philosopher Friedrich Nietzsche, a contemporary of Charles Darwin and an ardent evolutionist, popularized in Germany his concept of the superman, and then the master race. The ultimate outcome was Hitler, who elevated this philosophy to the status of a national policy.

In recent decades, the cause of racial liberation has made racism unpopular with intellectuals and only a few evolutionary scientists still openly espouse the idea of a long-term polyphyletic origin of the different races.[3] On the other hand, in very recent years, the pendulum has swung, and now we have highly vocal advocates of "black power" and "red power" and "yellow power," and these advocates are all doctrinaire evolutionists, who believe their own respective "races" are the fittest to survive in man's continuing struggle for existence.

According to the Biblical record of history, the Creator's divisions among men are linguistic and national divisions, not racial. Each nation has a distinct purpose and function in the corporate life of mankind, in the divine Plan (as, for that matter, does each individual).

"(God) hath made of one, all nations of men for to dwell on all the face of the earth, and hath determined the times before appointed, and the bounds of their habitation; That they should

---

1 Thomas Huxley: *Lay Sermons, Addresses and Reviews* (New York, Appleton, 1871) p. 20.

2 John S. Haller, Jr.: *Outcasts from Evolution — Scientific Attitudes of Racial Inferiority — 1850-1900.* (Urbana, University of Illinois Press, 1971), 228 pp.

3 One notable exception, among others, is the leading anthropologist Carleton Coon. See *The Origin of Races* (New York, Alfred Knopf, 1962), 724 pp.

seek the Lord, if haply they might feel after Him, and find Him" (Acts 17:26, 27).

No one nation is "better" than another, except in the sense of the blessings it has received from the Creator, perhaps in measure of its obedience to His Word and fulfillment of its calling. Such blessings are not an occasion for pride, but for gratitude.

Evolution and the Sexual Revolution

One of the most distressing developments on the modern scene is the breakdown of the institution of the family. In some states there are now as many divorces as there are marriages. A large proportion of troubled juveniles are known to come from broken homes. Even many families that manage to stay together seem to experience almost continual bickering, with no clearcut lines of authority and with low standards of moral behavior.

It is no mere coincidence that this modern deterioration of family life has occurred contemporaneously with the modern universal prevalence of evolutionary teaching. After all, God's creation and man's family life are closely associated in the Bible! The institution of marriage was the first human institution established by God and the command to have children was God's first commandment to man (Genesis 1:27, 28).

Because of the close relation of the home and family to God's creation, it is not surprising when we note today that a sound concept of marital and parental responsibilities goes hand-in-glove with a sound concept of Biblical creationism.

Similarly, it is no mere coincidence that the ascendancy of evolutionary philosophy in the past century was quickly followed by the decline of the sanctity of the home and marriage relationships. If man is not the special creation of God, then neither is the home. If man is an evolved animal, then the morals of the barnyard and the jungle are more "natural," and therefore more "healthy," than the artificially-imposed restrictions of premarital chastity and marital fidelity. Instead of monogamy, why not promiscuity and polygamy? Instead of training children in the nurture and admonition of the Lord, better to teach them how to struggle and survive in a cut-throat world, and then toss them out of the nest. Self-preservation is the first law of nature; only the fittest will survive. Be the cock-of-the-walk and the king-of-the-mountain! Eat, drink, and be merry, for life is short and that's the end. So says evolution!

Perhaps the greatest indictment of all against evolution is this assault against permanent, monogamous marriage and the sacred obligation of parents and children to each other. A strong emphasis on the full doctrine of Biblical creationism, in all its implications (including the proper Biblical roles of husband and wife) in both the home and church, is the best investment that can be made toward a happy home life, both in one's own home and in the future homes of one's children, and ultimately toward a healthy society and preparation for eternal responsibilities.

There are undoubtedly many personal and cultural reasons for the deterioration of family life in modern society and most of them can be shown to stem from the naturalistic, evolutionary philosophy which has been indoctrinated in young people for two generations or more through the schools and the media of mass communication. Perhaps the most important such factor involved in this breakdown has been the so-called "sexual revolution."

Pornography in almost every form is now freely available to all comers, not only in X-rated movies and newsstand paperbacks, but even on national television and in public school textbooks. People are constantly intimidated with "scientific" surveys which purportedly show majorities of both single and married people participating in pre-marital and extra-marital sex adventures, with the persuasive implication that what is done by "everyone" must be normal and therefore right for everyone.

The fact that sex outside of a permanent marriage bond is contrary to Scripture and to God's revealed will (note Hebrews 13:4; Ephesians 5:3-5; Matthew 19:3-9; etc.) is considered by evolutionists to be irrelevant, since the Bible is believed to be merely a product of man's religious evolution in an earlier stage of history and therefore no longer authoritative in our modern age of enlightenment and freedom.

Furthermore, since animals are indiscriminate with regard to partners in mating and, since men and women are believed to have evolved from animals, then why shouldn't we live like animals? Why develop sexual inhibitions and frustrations that may lead to psychological neuroses?

The modern psychological systems of Freud, Watson, Skinner, Rogers, and other leaders of the different schools of psychological thought today are all (whether Freudianism, behaviorism, humanistic psychology, or whatever may be the current fad in this field) based on the assumption that man is an animal, the product of ages of evolutionary struggle. On this assumption, people are counselled to release the sexual inhibitions that have been imposed on them by religion and act "naturally" (which, being interpreted, means to follow all their animal instincts) and engage in whatever sexual activity they desire with as many partners (of either sex) as they wish. Any unwanted children resulting from such activity can, of course, be taken care of either by abortion or by being taken over as wards of the state.

The concern expressed by Christian parents and pastors over the widespread introduction of sex education courses into the public schools is precisely because of the prominence of this kind of emphasis (sex as natural, with no moral connotations, based on the assumption of human evolution from an animal ancestry) in the texts and courses offered.

The sad testimony of multitudes of broken homes and broken lives, in contrast with the joyful testimony of multitudes of truly Christian families, is proof enough that evolutionary theory and the sexual revolution which has been based on it, is false and deadly. "A good tree cannot bring forth evil fruit, neither can a corrupt tree bring forth good fruit. Every tree that bringeth not forth good fruit is hewn down and cast into the fire" (Matthew 7:18, 19).

Evolution and Life in Outer Space

A very strange development of recent years has been the rapid growth of serious belief in the bizarre. Though there is not a shred of scientific evidence that biological life exists anywhere except on earth, both laymen and scientists in great numbers have come to believe that it does. Reports of unidentified flying objects and discoveries associated with the high technological skills of ancient men have been widely interpreted to mean that interplanetary astronauts from other worlds are frequent visitors to the earth.

The complete absence of evidence for such life in outer space is altogether ignored by those who believe in it. It is simply an article of faith that, since life has evolved on one planet in this solar system, it must also have evolved on planets in other solar systems. N.A.S.A. scientists, especially the geologists associated with this country's space program, have always admitted that the main purpose of the program was to find evidence as to how this solar system evolved and also to try to find proof of life in space.

This aspect of N.A.S.A.'s program, of course, proved to be a failure. There was no life on the moon nor on any other planet in the solar system, exactly as predicted by creationists all along.

Furthermore, there is no observational evidence of any other planets anywhere else in the universe! Astronomers believe there are millions of them, of course, but this idea is based squarely on evolutionary statistics! No one has ever actually *seen* any evidence of one, and science is supposed to be based on sight, not faith.

Astronomers have occasionally found evidence of molecules in space which are also found in living materials, and they have jumped to the conclusion that this somehow proves evolution has occurred "out there," as well as on earth. This conclusion is hardly justified by the evidence, however. Such molecules are immensely short of being *living* molecules! Furthermore, even these cannot exist very long and so could hardly evolve into higher molecules.

"Although it is clear that these molecules (e.g., OH, H_2O, NH_3, H_2CO, HCN, CN, HC_3N) exist in space and that they can emit radiation, there is no clear explanation of how they are formed and why they remain stable. Their chemical bonds should be broken by the intense fluxes of ultra-violet radiation and cosmic rays. Their estimated lifetimes from ultraviolet dissociation in interstellar space is about 200 years."[1]

---

[1] Gerald L. Wick: "Interstellar Molecules: Chemicals in the Sky," *Science,* Vol. 170, October 9, 1970, p. 1295.

Speculation seems to be unhindered by such facts, however. Many eminent scientists continue to believe that life *must* be out there somewhere because it would be improbable that it would evolve only once in the entire universe.

"They hypothesize that thousands of millions of years ago, an intelligent civilization decided to seed other nearby planets with primitive forms of life in the hope that more advanced civilizations might develop. Crick and Orgel claim that their proposal — called Directed Panspermia — is as tenable as other theories that aim to explain the origin of life on earth."[1]

If that be so, why then do we not find *some* evidence of intelligent life in other parts of the universe? A remarkable new theory suggests that this is because the celestial astronauts do not want us to know they are there!

"A humble explanation of the null results of the searches for life is advanced by a Harvard student, John Ball (*Icarus,* Vol. 19, p. 347). He steps into the realms of science fiction with a hypothesis that we are living in a galactic zoo! The idea is that a super civilization may by now have control of the whole Galaxy. Just as we have Safari Parks, zoos and conservation areas, so they may have set aside the solar system as a wilderness zone. The perfect zoo keeper does not make himself known to his charges, and thus we are unaware of their presence."[2]

Since there is no evidence of life in outer space, so the argument goes, therefore it must be there! This kind of evidence is, to say the least, not compelling evidence! Nevertheless if men do not believe in a divine Creator, they must eventually believe in some kind of spontaneous generation of life. Such an idea as that, of course, clearly contradicts the scientific law of cause-and-effect.

As far as the evidence of high technological skills found in ancient structures and drawings is concerned, this in no way proves that these skills were imported from somewhere in outer space.

---

[1] Simar, Mitton and Roger Lewin: "Is Anyone Out There?" *New Scientist,* August 16, 1973, p. 380.

[2] *Ibid,* p. 382.

To the extent that such evidences are genuine, they can be much more easily and directly explained in terms of the creation model, which suggests that the earliest men, being highly intelligent and living to great ages, had developed a very high civilization, both before the great Flood and, at least in many places, again quite soon after the Flood.

The creation model alone has a scientifically defensible explanation for life. It was specially created by a living Creator, who, Himself, has existed from eternity. This explanation is perfectly consistent with causality and all *known* data. And, as far as we can tell, biological life is unique to this earth.

Evolution and the Public Schools

The current revival of interest in creationism has focused largely on the public schools. Parents, pastors and others have become painfully aware that, in the name of "science," their young people are being thoroughly indoctrinated in the deadly philosophy of evolutionary humanism. Biblical creationism is either ridiculed or (which is even worse) ignored as not worth mentioning.

This indoctrination is accomplished not only formally and directly in, say, biology and geology courses, but even more effectively through indirect application of evolutionary assumptions in the social sciences and humanities, as well as in the very methodology of modern educationism. Occasionally an educator will react defensively to such charges by insisting that, in his class or school, evolution is presented only as a "theory" and that other theories are also included. What this usually means is that various mythological ideas of origins are mentioned rapidly in passing (Hindu, American Indian, Babylonian, etc., as well as the "story" in Genesis) and then the rest of the semester is spent in studying the only "theory" which is taken seriously by *scientists* — namely evolution!

One of the leading evolutionists of our times, Dr. Theodosius Dobzhansky, made a fascinating comment and admission in one of his innumerable articles several years ago:

"It would be wrong to say that the biological theory of evolution has gained universal acceptance among biologists or even among geneticists. This is perhaps unlikely to be achieved by any theory which is so extraordinarily rich in philosophic and humanistic implications. Its acceptance is nevertheless so wide that its opponents complain of inability to get a hearing for their views."[1]

Note that evolution is "rich in philosophic and humanistic implications." It is those very implications to which creationists object. It is also this fact of an implied philosophy of humanism that brands evolution as fundamentally a *religion,* rather than a science!

When creationists propose, however, that creation be taught in the schools along with evolution, evolutionists commonly react emotionally, rather than scientifically. Their "religion" of naturalism and humanism has been in effect the established religion of the state for a hundred years, and they fear competition. They usually refuse to teach scientific creationism on the basis of an obviously false claim that creation is religion and evolution is science.

In the ultimate sense, no concept of origins can really be scientific. In the present world, neither evolution nor creation is taking place, so far as can be observed (and science is supposed to be based on observation!). Cats beget cats and fruit-flies beget fruit-flies. Life comes only from life. There is nothing new under the sun.

Neither evolution nor creation is accessible to the scientific method, since they deal with origins and history, not with presently observable and repeatable events. They can, however, be formulated as scientific *models,* or frameworks, within which to predict and correlate observed facts. Neither can be *proved;* neither can be *tested.* They can only be *compared* in terms of the relative ease with which they can explain data which exist in the real world.

There are, therefore, sound scientific and pedagogical reasons why *both* models should be taught, as objectively as possible, in public classrooms, giving arguments pro and con for each. Some

---

[1] Theodosius Dobzhansky: "Evolutionary and Population Genetics, *Science,* Vol. 142, November 29, 1963, p. 1134.

students and their parents believe in creation, some in evolution, and some are undecided. If creationists desire *only* the creation model to be taught, they should send their children to private schools which do this; if evolutionists want only evolution to be taught, they should provide private schools for *that* purpose. The public schools should be neutral and either teach both or teach neither.

This is clearly the most equitable and constitutional approach. Many people have been led to believe, however, that court decisions restricting "religious" teaching in the public schools apply to "creation" teaching and not to "evolution" teaching. Nevertheless, creationism is actually a far more effective scientific model than evolutionism, and evolution requires a far more credulous religious faith in the illogical and unprovable than does creation. An abundance of sound scientific literature is available today to document this statement, but few evolutionists have bothered to read any of it. Many of those who *have* read it have become creationists!

What can creationists do to help bring about a more equitable treatment of this vital issue in the public schools? How can they help their own children in the meantime? The following suggestions are in order of recommended priority. All involve effort and expense, and the need is urgent.

(1) Most basic is the necessity for each concerned creationist himself to become informed on the issue and the scientific facts involved. He does not need to be a scientist to do this, but merely to read several of the scholarly creationist books that are now available. He should also study creationist literature that demonstrates the fallacious nature of the various compromising positions (e.g., theistic evolution, day-age theory, gap theory, local flood theory, etc.) in order to be on solid ground in his own convictions.

(2) He should then see that his own children and young people, as well as others for whom he is concerned, have access to similar literature on their own level. He also should be aware of the teachings they are currently receiving in school and help them find answers to the problems they are encountering. He should encourage them always to be

gracious and respectful to the teacher, but also to look for opportunities (in speeches, term papers, quizzes, etc.) to show that, although they understand the arguments for evolution, the creationist model can also be held and presented scientifically.

(3) If he learns of teachers who are obviously bigoted and unfair toward students of creationist convictions, it would be well for him to talk with the teacher himself, as graciously as possible, pointing out the true nature of the issue and requesting the teacher to present both points of view to the students. Under some circumstances, this might be followed up by similar talks with the principal and superintendent.

(4) Many teachers and administrators are quite willing to present both viewpoints, but have been unaware that there does exist a solid scientific case for creation, and, therefore, they don't know how to do this. There is thus a great need for teachers, room libraries, and school libraries to be supplied with sound creationist literature. Perhaps some schools, or even districts, will be willing to provide such literature themselves. If not, the other alternative is for parental associations, churches, or individuals to take on such a project as a public service. If sound creationist books are conveniently available, many teachers (not all, unfortunately, but far more than at present) would be willing to use them and to encourage their students to use them.

(5) Creationist parents, teachers, pastors, and others can join forces to sponsor meetings, seminars, teaching institutes, etc., in their localities. Qualified creationist scientists can be invited to speak at such meetings, and if adequate publicity (especially on a person-to-person basis) is given, a real community-wide impact can be made in this way. Especially valuable, when such invitations can be arranged, are opportunities for creationist scientists to speak at meetings of scientists or educators. Also such men can be invited to speak in churches or in other large gatherings of interested laymen.

(6) Discussions can be held with officials at high levels (state education boards, district boards, superintendents, etc.) to acquaint them with the evidences supporting creation and the importance of the issue. They can be requested to inform the teachers of their state or district that the equal teaching of evolution and creation, not on a religious basis, but as scientific models, is both permitted and encouraged. Cases of unfair discrimination against creationist minorities in classrooms can be reported, and most officials at such levels are sufficiently concerned with the needs of *all* their constituents that, if they can first be shown there is a valid scientific case for creation and that evolution has at least as much religious character as does creation, they will quite probably favor such a request.

(7) Public response can be made (always of a scientific, rather than emotional flavor) to newspaper stories, television programs, etc., which favor evolution. Those responses may be in the form of letters-to-the-editor, protest letters to sponsors, news releases, and other means.

(8) Financial support should be provided for those organizations attempting in a systematic way to do scientific research, produce creationist textbooks and other literature, and to provide formal instruction from qualified scientists in the field of creationism. This can be done both through individual gifts and bequests and through budgeted giving by churches and other organizations.

It will be noted that no recommendation is made for political or legal pressure to *force* the teaching of creationism in the schools. Some well-meaning people have tried this, and it may serve the purpose of generating publicity for the creationist movement. In general, however, such pressures are self-defeating. "A man convinced against his will is of the same opinion still."

Force generates reaction, and this is especially true in such a sensitive and vital area as this. The hatchet job accomplished on the fundamentalists by the news media and the educational establishment following the Scopes Trial in 1925 is a type of what could happen, in the unlikely event that favorable legislation or court decisions could be obtained by this route.

Reasonable persuasion is the better route. "The servant of the Lord must not strive, but be gentle unto all men, apt to teach, patient, In meekness instructing those that oppose themselves"(II Timothy 2:24, 25).

As far as the students themselves are concerned, what should be their attitudes and actions? How should a student who is a Christian and creationist behave in a class taught by a teacher who is a strong and opinionated evolutionist?

If the student is silent about his convictions and pretends to go along with the classroom teachings, is this a hypocritical compromise for the sake of expediency? On the other hand, if he challenges the teacher, arguing and taking an open stand against the evolutionary philosophy, will this not result in a failing grade in the course, ridicule by the teacher and other students, and possibly even shut the door to the career he has chosen?

This is a very real problem, with no easy answer. Essentially the same question is asked also by college students and even by graduate students working on their Ph.D. degrees. The writer has known students who have failed courses, and some who have been denied admission to graduate school or have been hindered from obtaining their degree, largely for this very reason. When the writer was on the faculty at Virginia Tech, a professor who was on the graduate faculty there in the Biology Department said that he would never approve a Ph.D. degree for any student known to be a creationist in his department, even if that student made straight A's in all his courses, turned in an outstanding research dissertation for his Ph.D., and was thoroughly familiar with all the evidences and arguments for evolution.

On the other hand, there have been many other students who have received excellent grades even while taking a strong stand against evolution, as well as many who have been awarded graduate degrees, in spite of being known as Christians with solid creationist convictions.

What makes the difference? There is certainly no simplistic solution, applicable always and everywhere. Individual teachers are different, schools are different, and students are different, and these differences all make a difference! However, there are certain general principles that should always be at least considered:

(1) Wherever possible, one should bypass the problem by enrolling in a Christian school, or at least in a class with a Christian teacher. Whenever such vital questions as origins or basic meanings are to be discussed in courses or textbooks, the happiest situation is for both student and teacher to have the same ultimate motives and goals. Unfortunately, this solution is often impossible or impracticable.

(2) As long as a student is enrolled in a given class or program, he is under the authority of the teacher and is there for the primary purpose of learning, rather than witnessing. He should, therefore, at all times be respectful and appreciative, doing his best to learn the material presented, whether he agrees with the teacher's personal philosophy or not. This is the Biblical admonition (Titus 2:9, 10; Colossians 3:22-24; Ephesians 6:5-8) concerning masters and servants, and this would apply in principle at least to the relation between (school) masters and those in their charge. Also, especially in the case of minor children, the teacher is *in loco parentis,* and the obedience of children as to parents is commanded (Galatians 4:1, 2). If this situation becomes intolerable, due to gross irresponsibility or abuse of authority on the teacher's part, then probably the proper course is to withdraw from the class, giving a careful and objective explanation in writing and with documentation, of the reasons for withdrawal, to both the teacher and administrator concerned.

(3) Differences in attitudes and beliefs between teachers and students can often be resolved, or at least ameliorated, by a sincere attempt to maintain an attitude of objectiveness and good humor relative to their differences. Emotional arguments, especially when defensively oriented around religious convictions, will alienate, rather than attract (Proverbs 15:1; 25:11; Colossians 4:6).

(4) The student should be well-informed on both sides of the evolution-creation question, so that such objections as he may have opportunity to raise (whether in class, on term

papers, in formal debates, or by other means) will be based on sound evidence, not on hearsay or misunderstanding. Most teachers (not all, unfortunately, but most) will respond with interest and fairness to a well-prepared and soundly-reasoned argument for creationism, especially if presented objectively and scientifically, in an attitude of respect and good will (II Timothy 2:15; I Peter 3:15).

(5) Other things being equal, a person should be able to do a better job in any course or at any task if he is a Christian than he could have before becoming a Christian, since he now has greater resources and higher motives than before. The subject matter of any course has value to him as a Christian witness, even if for nothing else than to make him better informed concerning what others believe. Therefore, he should study diligently and do the best job of which he is capable (I Corinthians 10:31; Colossians 3:23). Any teacher is more likely to respond favorably to the suggestions of a good student than of a poor, lazy, belligerent student.

(6) Finally, there is no substitute for a consistent and winsome Christian walk in public and a life based on prayer and the study of Scripture in private, in meeting this particular problem as well as other problems involving similar tensions and confrontations in life (Proverbs 16:7).

Evolution Undermining Christian Education

The evolutionary philosophy thoroughly dominates the curricula and faculties of secular colleges and universities today, as most people are well aware. It is not so well known, however, that this philosophy has also had considerable effect on many Christian colleges. When this fact is pointed out, the reaction of many Christians seems to be one of surprise or even doubt. "How could . . . College, so well known for academic leadership in the Christian world, possibly be teaching evolution, especially when its faculty members all assent to a statement of faith? Surely there must be some mistake."

There is no mistake, however. Although there are still many Christian schools whose faculties are strongly Biblical and strictly creationist, many of the most highly respected schools have compromised with evolutionism to an alarming degree. A recent article entitled "Creationism and Evolutionism as Viewed in Consortium Colleges" (*Universitas,* Vol. 2, No. 1, March 1974) documents this fact quite thoroughly. Written by Dr. Albert J. Smith, a biology teacher at Wheaton College, this paper gives the views of 38 teachers in science and math from the Christian College Consortium, a group of about 10 or 12 of the leading Christian colleges, including such schools as Wheaton College, Gordon College, Westmont College, Messiah College, Malone College, Taylor University, Seattle Pacific College, Greenville College, Bethel College and Eastern Mennonite College.

Dr. Smith points out that, in the opinion of their own science faculties, these institutions have *no well-defined position* on creation or evolution. Nevertheless, these people also say that their institutions must "maintain a conservative stance for promotional purposes." Interesting! Financial supporters of Christian schools are usually strong creationists.

As far as the faculty members themselves were concerned, Dr. Smith says: "Efforts to characterize and identify with the departmental positions results in all respondents calling themselves 'theistic evolutionists,' 'progressive creationists,' or infrequently 'fiat creationists.' "

It is good to know there are still a few "fiat creationists" in the Consortium, but it is evident they constitute a small minority. "Progressive creationism," of course, is a semantic variant of "theistic evolutionism," both systems adopting the geologic-age framework which is essentially synonymous with naturalistic uniformitarianism and rejecting the straightforward Biblical teaching of a completed recent creation and worldwide flood.

None of the colleges in the Consortium openly teach evolutionism in the manner of secular colleges, of course. Some teachers do try to present both creation and evolution, and the evidences for and against each, to their students. The predominant attitude, however, is apparently that the question of origins is unimportant and irrelevant. "Relatively few colleges emphasize the creationist-evolutionist dialogue at all. . . . The students are encouraged to make up their own minds regarding personal position."

Quotations given in the article from the individual responses of faculty members show that many of them use the standard cliches in trying to avoid this question. ". . . creationism (a Biblical statement) and evolutionism (a scientific statement) are not considered to be antagonistic but rather at different levels; creationism considers the *who* and the *why* while evolution considers the *how,* the *when,* and the *how much.*" " . . . the important thing is not *how* but *Who.*"

The reason why true creationists object to such views, of course, is because the Bible *does* say how, when, and how much, as well as who and why! Furthermore, as the scientists of the Creation Research Society and the Institute for Creation Research have shown, the true facts of science do correlate much better with these Biblical statements on creation than with evolution. Such facts, however, seem always to be ignored by "Christian evolutionists" and "progressive creationists."

Saying that the evolution/creation question is "not a significant problem," "not basic to the Christian faith," "unimportant," "a dead issue," and the like (all these judgments are quoted from respondents of the Consortium) is most likely merely a devious way of saying: "Well, acceptance in the academic world requires me to believe in evolution, and I don't want to face up to the Biblical and scientific reasons for rejecting evolution, so I would prefer to bypass the problem."

On one occasion several years ago, the writer spent several hours discussing this problem with a professor of geology of one of the Consortium colleges. This teacher insisted that Christians *must* accept the geological-age system as taught by evolutionary geologists. When asked how he, as professedly a Bible-believing Christian, reconciled the Genesis record of creation and the flood with this system, his reply was that he didn't know of any way in which they *could* be reconciled (he agreed that neither the gap theory nor the day-age theory was acceptable). When also asked how he reconciled Jesus Christ's acceptance of the literal Genesis record of creation and the flood with the geological ages, he replied that he didn't know how to reconcile that either. His final conclusion was that all of this was unimportant anyway. Only one thing apparently *was* important; namely, to accept the geological ages!

More recently, the writer had two opportunities to talk at some length with the present Head of the Geology Department at this same college on the same subject. He took much the same position, also adding that we would never be able to understand the meaning of the Genesis record of creation and the flood until we get to heaven! It is not important for us to understand it now, he felt.

One must confess a certain lack of patience with this type of logic. How can a Christian say the doctrine of special creation is unimportant when it is foundational to every other doctrine in Scripture? How can one say the evolutionary philosophy is not significant, when it has been made the basis of fascism, communism, animalism, racism, modernism, atheism, and practically every other harmful philosophy known to man? How can *Christian* college professors teach their students that evolution is an optional question when the Scriptures plainly teach otherwise?

"How long halt ye between two opinions? If the Lord be God, follow him: but if Baal, then follow him" (I Kings 18:21).

Although evolutionism has most affected Christian higher education, this philosophy has also influenced numerous Christian elementary and secondary schools. Many leading Christian periodicals (*Eternity, Christianity Today, Christian Life,* etc.) have likewise been significantly infiltrated by evolutionary thinking. Many churches, missions, and other Christian institutions have been similarly affected.

One of the most frustrating problems encountered in trying to encourage and strengthen belief in a Creator and in creationism is the indifference of so many professing Christian people to the urgent importance of this issue. "I don't believe in evolution anyhow, so why should I waste time in studying or promoting creationism?" "Why get involved in peripheral and controversial issues like that — just preach the Gospel!" "The Bible is not a textbook of science, but of how to live." "It is the Rock of Ages which is important — not the age of rocks!" "Winning souls is the principal thing — not the winning of debates."

Platitudes such as the above, however spiritual they sound, are really cop-outs. They tend to become excuses for avoiding serious thought and the offense of the cross. In the name of evangelism and of appealing to large numbers, a least-common-denominator emphasis on emotional experiences and a nominal commitment of some kind has become the dominant characteristic of most Christian teaching and activity today, and this is almost as true in fundamentalist and conservative circles as it is among religious liberals.

This ostrich-like attitude seems to date largely from the after-effects of the infamous Scopes Trial in 1925. The fundamentalists and creationists were made to look so ridiculous by the news media covering that trial at the time (and they still are exploiting it today!) that Christians in general retreated altogether from the battle for the schools and the minds of the young people. Avoiding any further attempt to relate science and history to an inspired Bible, Christian teachers and preachers thenceforth emphasized evangelism and the spiritual life almost exclusively. The "gap theory" which supposedly allowed the earth's billions of years of evolutionary history to be pigeon-holed between the first two verses of Genesis and then ignored, provided a convenient device for saying the entire question was irrelevant.

As a consequence, in less than a generation, the entire school system and the very establishment itself — educational, scientific, political, military, industrial, and religious — was taken over by the evolutionary philosophy and its fruits of naturalism, humanism, socialism, and animalistic amoralism.

For the past decade, however, a noteworthy revival of creationism has been taking place, both in the churches and, to some extent, in the schools. Thousands of scientists have become creationists, and the interest among teachers and students in creationism is higher than it has ever been.

Nevertheless, although many churches and Christian people have become actively involved in the creation issue, it is still sadly true that the majority of them are indifferent, or even antagonistic, to creationism. They think it is only a peripheral biological question, of no concern in the preaching of the Gospel. Even most fundamentalists, who themselves believe in creation, think evolution is a dead issue.

Such an attitude is based on wishful thinking, to say the least. The lead article in a recent issue of *Science,* the official journal of the prestigious American Association for the Advancement of Science, says:

"While many details remain unknown, the grand design of biologic structure and function in plants and animals, including men, admits to no other explanation than that of evolution. Man, therefore is another link in a chain which unites all life on this planet."[1]

Not only did man evolve, but so did "the religions of Jesus and Buddha."[2] That being so, not only are the supernatural aspects of Christianity open to question, but so are its ethical teachings.

"An ethical system that bases its premises on absolute pronouncements will not usually be acceptable to those who view human nature by evolutionary criteria."[3]

Ethics and morals must evolve as well as organisms! And so must social and political systems. There are no absolutes.

This is the logical and inevitable outgrowth of evolutionary teaching. This is also the logical and inevitable outgrowth of Christian indifference to evolutionary teaching.

The doctrine of special creation is the foundation of all other Christian doctrine. The experience of belief in Christ as Creator is the basis of all other Christian experience. Creationism is not peripheral or optional; it is central and vital. That is why God placed the account of creation at the beginning of the Bible, and why the very first verse of the Bible speaks of the creation of the physical universe.

Jesus Christ was Creator (Colossians 1:16) before He became Redeemer (Colossians 1:20). He is the very "beginning of the creation of God" (Revelation 3:14). How then can it be possible to really know Him as Saviour unless one also, and first, knows God as Creator?

The very structure of man's time commemorates over and over

---

[1] A. G. Motulsky: "Brave New World," *Science,* Vol. 185, August 23, 1974, p. 653.

[2] *Ibid.*

[3] *Ibid,* p. 654.

again, week by week, the completed creation of all things in six days. The preaching of the Gospel necessarily includes the preaching of creation. ". . . the everlasting gospel to preach unto them that dwell on the earth . . . worship Him that made heaven, and earth, and the sea, and the fountains of waters" (Revelation 14:6, 7).

If man is a product of evolution, he is not a fallen creature in need of a Saviour, but a rising creature, capable of saving himself.

"The ethical human brain is the highest accomplishment of biologic evolution." [1]

The gospel of evolution is the enemy of the Gospel of Christ. The Gospel of Christ leads to salvation, righteousness, joy, peace, and meaning in life. Evolution's gospel yields materialism, collectivism, anarchism, atheism, and despair in death.

Evolutionary thinking dominates our schools today — our news media, our entertainment, our politics, our entire lives. But evolution is false and absurd scientifically! How long will Christian people and churches remain ignorant and apathetic concerning it?

Evolution versus the Bible

The evolutionary system has been entrenched for so long that many people who otherwise accept the Bible as infallible have deemed it expedient to compromise on this issue. Thus, evolution has been called, "God's method of creation," and the Genesis record of the six days of creation has been reinterpreted in terms of the evolutionary ages of historical geology. These geological ages themselves have been accommodated in Genesis either by placing them in an assumed "gap" between Genesis 1:1 and 1:2 or by changing the "days" of creation into the "ages" of evolution.

Theories of this kind raise more problems than they solve, however. It is more productive to take the Bible literally and then to interpret the actual facts of science within its revelatory framework. If the Bible cannot be understood, it is useless as

---

1 Motulsky, *op cit,* p. 662.

revelation. If it contains scientific fallacies, it could not have been given by *inspiration.*

The specific purpose of this final section is to show that all such theories which seek to accommodate the Bible to evolutionary geology are invalid and therefore should be abandoned.

Evolution is believed by its leading advocates to be a basic principle of continual development, of increasing order and complexity, throughout the universe. The complex elements are said to have developed from simpler elements, living organisms to have evolved from non-living chemicals, complex forms of life from simpler organisms, and even man himself to have gradually evolved from some kind of ape-like ancestor. Religions, cultures, and other social institutions are likewise believed to be continually evolving into higher forms.

Thus, evolution is a complete world-view, an explanation of origins and meanings without the necessity of a personal God who created and upholds all things. Since this philosophy is so widely and persuasively taught in our schools, Christians are often tempted to accept the compromise position of "theistic evolution," according to which evolution is viewed as God's method of creation. However, this is basically an inconsistent and contradictory position. A few of its fallacies are as follows:

(1) It contradicts the Biblical record of creation. Ten times in the first chapter of Genesis, it is said that God created plants and animals to reproduce "after their kinds." The Biblical "kind" may be broader than our modern "species" concept, but at least it implies definite limits to variation. The New Testament writers accepted the full historicity of the Genesis account of creation. Even Christ Himself quoted from it as historically accurate and authoritative (Matthew 19: 4-6).

(2) It is inconsistent with God's methods. The standard concept of evolution involves the development of innumerable misfits and extinctions, useless and even harmful organisms. If this is God's "method of creation," it is strange that He would use such cruel, haphazard, inefficient, wasteful processes. Furthermore, the idea of the "survival of the fittest," whereby the stronger animals eliminate the

weaker in the "struggle for existence" is the essence of Darwin's theory of evolution by natural selection, and this whole scheme is flatly contradicted by the Biblical doctrine of love, of unselfish sacrifice, and of Christian charity. The God of the Bible is a God of order and of grace, not a God of confusion and cruelty.

(3) The evolutionary philosophy is the intellectual basis of all anti-theistic systems. It served Hitler as the rationale for Nazism and Marx as the supposed scientific basis for communism. It is the basis of the various modern methods of psychology and sociology that treat man merely as a higher animal and which have led to the mis-named "new morality" and ethical relativism. It has provided the pseudo-scientific rationale for racism and military aggression. Its whole effect on the world and mankind has been harmful and degrading. Jesus said: "A good tree cannot bring forth evil fruit" (Matthew 7:18). The evil fruit of the evolutionary philosophy is evidence enough of its evil roots.

Thus, evolution is Biblically unsound, theologically contradictory, and sociologically harmful.

Some Christians use the term "progressive creation" instead of "theistic evolution," the difference being the suggestion that God interjected occasional acts of creation at critical points throughout the geological ages. Thus, for example, man's soul was created, though his body evolved from an ape-like ancestor.

This concept is less acceptable than theistic evolution, however. It not only charges God with waste and cruelty (through its commitment to the geologic ages) but also with ignorance and impotence. God's postulated intermittent creative efforts show either that He didn't know what He wanted when He started the process or else that He couldn't provide it with enough energy to sustain it until it reached its goal. A god who would have to create man by any such cut-and-try, discontinuous, injurious method as this can hardly be the omniscient, omnipotent, loving God of the Bible.

According to the established system of historical geology, the history of the earth is divided into a number of geological ages. The earth is supposed to have evolved into its present form and inhabitants over a vast span of geologic ages, beginning about five billion years ago.

In contrast, the Biblical revelation tells us that God created the entire universe in six days only a few thousand years ago. Consequently, many Christian scholars have tried to find some way of reinterpreting Genesis to fit the framework of earth history prescribed by the geologists.

The most popular of these devices has been the "day-age" theory, by which the "days" of creation were interpreted figuratively as the "ages" of geology. However, there are many serious difficulties with this theory.

The Hebrew word for "day" is *yom,* and this word can occasionally be used to mean an indefinite period of time, if the context warrants. In the overwhelming preponderance of its occurrences in the Old Testament, however, it means a literal day; that is, either an entire solar day or the daylight portion of a solar day. It was, in fact, defined by God Himself the very first time it was used, in Genesis 1:5, where we are told that "God called the light, day." It thus means, in the context, the "day" in the succession of "day and night" or "light and darkness."

Furthermore, the word is never used to mean a definite period of time, in a succession of similar periods (that is, "the first day," "the second day," etc.) or with definite terminal points (that is, as noted by "evening and morning," etc.) unless that period is a literal solar day. And there are hundreds of instances of this sort in the Bible.

Still further, the plural form of the word (Hebrew *yamim*) is used over 700 times in the Old Testament and clearly refers to literal days. There are a few passages of uncertain meaning, but even in these the literal interpretation is most probable. Never can it be demonstrated that *yamim* was intended by the writer to mean anything except natural solar days. A statement in the Ten Commandments, written on a tablet of stone directly by God Himself, is very significant in this connection, where He uses this word and says plainly: "In six days, the Lord made heaven and earth, the sea, and all that in them is" (Exodus 20:11).

Not only is the day-age theory unacceptable Scripturally, but it also is grossly in conflict with the geological position with which it attempts to compromise. There are more than 20 serious contradictions between the Biblical order and events of the creative days and the standard geologic history of the earth and its development, even if it were permissible to interpret the "days" as "ages." For example, the Bible teaches that the earth existed before the stars, that it was initially covered by water, that fruit trees appeared before fishes, that plant life preceded the sun, that the first animals created were the whales, that birds were made before insects, that man was made before woman, and many other such things, all of which are explicitly contradicted by historical geologists and paleontologists.

But the most serious fallacy in the day-age theory is theological. It charges God with the direct responsibility for five billion years of history of purposeless variation, accidental changes, evolutionary blind alleys, numerous misfits and extinctions, a cruel struggle for existence, with preservation of the strong and extermination of the weak, of natural disasters of all kinds, rampant disease, disorder and decay, and, above all, with death. The Bible teaches that, at the end of the creation period, God pronounced His whole creation to be "very good," in spite of all this. It also teaches plainly that this present type of world, "groaning and travailing in pain" (Romans 8:22) only resulted from man's sin and God's curse thereon. "By one man sin entered into the world, and death by sin" (Romans 5:12). "God is not the author of confusion" (I Corinthians 14:33).

There are two possibilities for theories for harmonizing the first chapter of Genesis with the geologic ages. One places the geologic ages "during" the six days of creation (thus making the "days" into "ages"), and the other places the geologic ages "before" the six days (thus making them days of "re-creation" following a great cataclysm which had destroyed the primeval earth). The "day-age theory" has been shown to be an impossible compromise, both Biblically and scientifically.

The "gap theory" likewise involves numerous serious fallacies. The geologic ages cannot be disposed of merely by ignoring the extensive fossil record on which they are based. These supposed ages are inextricably involved in the entire structure of the evolutionary history of the earth and its inhabitants, up to and including man. The fossil record is the main argument for evolution (in fact, the only such argument which suggests evolution on more than a trivial scale). Furthermore, the geologic ages are recognized and identified specifically by the fossil contents of the sedimentary rocks in the earth's crust. The very names of the ages show this. Thus, the "Paleozoic Era" is the era of "ancient life," the "Mesozoic Era" of "intermediate life," and the "Cenozoic Era" of "recent life." As a matter of fact, the one primary means for dating these rocks in the first place has always been the supposed "stage-of-evolution" of the contained fossils.

Thus, acceptance of the geologic ages implicitly involves acceptance of the whole evolutionary package. Most of the fossil forms preserved in the sedimentary rocks have obvious relatives in the present world, so that the "re-creation" concept involves the Creator in "re-creating" in six days many of the same animals and plants which had been previously developed slowly over long ages, only to perish violently in a great pre-Adamic cataclysm.

The gap theory, therefore, really does not face the evolution issue at all, but merely pigeon-holes it in an imaginary gap between Genesis 1:1 and 1:2. It leaves unanswered the serious problem as to why God would use the method of slow evolution over long ages in the primeval world, then destroy it, and then use the method of special creation to re-create the same forms He had just destroyed.

Furthermore, there is no geologic evidence of such a worldwide cataclysm in recent geologic history. In fact, the very concept of a worldwide cataclysm precludes the geologic ages, which are based specifically on the assumption that there have been no such worldwide cataclysms. As a device for harmonizing Genesis with geology, the gap theory is self-defeating.

The greatest problem with the theory is, again, that it makes God the direct author of evil. It implies that He used the methods of struggle, violence, decay, and death on a worldwide scale for at least three billion years in order to accomplish His unknown purposes in the primeval world. This is the testimony of the fossils and the geologic ages which the theory tries to place before

Genesis 1:2. Then, according to the theory, Satan sinned against God in heaven (Isaiah 14:12-15; Ezekiel 28:11-17), and God cast him out of heaven to the earth, destroying the earth in the process in the supposed pre-Adamic cataclysm. Satan's sin in heaven, however, cannot in any way account for the age-long spectacle of suffering and death in the world during the geologic ages which *preceded* his sin! Thus, God alone remains responsible for suffering, death, and confusion, and without any reason for it.

The Scripture says, on the other hand, at the end of the six days of creation, "And God saw everything that he had made (e.g., including not only the entire earth and all its contents, but all the heavens as well — note Genesis 1:16; 2:2, etc.) and, behold, it was very good" (Genesis 1:31). Death did not "enter the world" until man sinned (Romans 5:12; I Corinthians 15:21). Evidently even Satan's rebellion in heaven had not yet taken place, because everything was pronounced "very good" there, too.

The real answer to the meaning of the great terrestrial graveyard — the fossil contents of the great beds of hardened sediments all over the world — will be found neither in the slow operation of uniform natural processes over vast ages of time nor in an *imaginary* cataclysm that took place before the six days of God's perfect creation. Rather, it will be found in a careful study of the very *real* worldwide cataclysm described in Genesis 6 through 9 and confirmed in many other parts of the Bible and in the early records of nations and tribes all over the world; namely, the great Flood of the days of Noah.

Only a few of the many difficulties with the various accommodationist theories have been discussed, but even these have shown that it is impossible to devise a legitimate means of harmonizing the Bible with evolution. We must conclude, therefore, that if the Bible is really the Word of God (as its writers allege and as we believe) then evolution and its geological age-system must be completely false.

Let God Be True

We have now traced the deep roots of evolutionism, and seen some of its bitter fruits. Its origin is far back in pagan antiquity, with its tentacles permeating all the religions and philosophies of early man. Its reimposition on the modern world a century ago was accomplished by a strange complex of pressures and propaganda, promoted in the name of-science but completely void of any real scientific evidence.

Since that time, its applications in social and political thinking have been mainly responsible for the vast movements of economic and military imperialism, for the spreading influence of communism, for many forms of virulent racism, for anarchism, for the various animalistic psychologies, and for the growing cancer of practical atheism and materialism in all aspects of modern life.

With the superficially attractive semantics of humanism as its garb, the evolutionary philosophy has now captured our entire educational system. The doctrine of a personal God and true creationism, as well as all other aspects of Biblical Christianity (and orthodox Judaism as well) have been banned from the schools because of their religious connotations, while another religion — that of evolutionary pantheistic humanism — has been universally substituted for it in the name of science. As we have seen, this philosophy is now foundational in the teaching of all the natural and social sciences, as well as in the humanities and fine arts, and it has been woven into the very presuppositions and methodology of the entire educational enterprise. Not only so, but the communications media, the political system, the scientific establishment, the military-industrial complex, and most other components of modern society, now seem to worship, consciously or unconsciously, at the shrine of the great god Evolution!

We have further noted that, having generated most of the world's modern problems — pollution, energy depletion, racism, political revolutions, global wars, educational deterioration, moral disintegration, etc. — not to mention communism — the evolutionary philosophy is now making the problems still worse by attempting to solve them in the framework of its own false premises.

It is no wonder, therefore, that many people have become disillusioned with the whole evolutionary system and that a serious revival of creationism is beginning to take form. Millions of intelligent and serious laymen, including thousands of qualified scientists, are now calling for a return to faith in God as Creator and in the Bible as the true record of His works of creation and redemption.

There is no more important question for a person to face than this: "Am I only the chance product of forces of nature operating randomly over aeons of time, with no real meaning and no real future, or am I a special creation of an omnipotent, loving personal God?"

When one seriously seeks the answer to this question, he will find that the real facts of science and history support creation, even though the leaders of the scientific and educational establishments insist on evolution. God *did* create the world, whether men believe it or not, and the evidence in the real world is bound to agree with this. Furthermore, God *is* going to return someday to judge the world, skeptics to the contrary notwithstanding.

Obviously, no one can prove there is no Creator, even though he may not wish to believe in Him. Therefore, each person should at least make a careful examination of the main lines of evidence that bear on the question of origins. He owes it to himself to do this, as well as to other people whose beliefs and attitudes he may influence.

If, indeed, there *is* a Creator (and there is much stronger evidence for it than against it), it is important that we base our plans and decisions on this fact. A divine Creator, by definition, is omniscient, and therefore knows all about each of us. A Creator must be omnipotent, and therefore able to control the world and all its creatures. Furthermore, the Creator, again by definition, is purposeful and not capricious, which means He has a purpose for this world and for each of our individual lives. He has good reason for everything He causes to happen, or even allows to happen. He even has good reason for letting people decide for themselves whether to believe in Him or not. He is a *loving* God, and thus wants everyone to know and love Him but, by the same token, He will not *force* anyone to do so. He has provided a tremendous body of evidence to encourage our faith in Him, but not so much as to make it impossible *not* to believe, if that is what one prefers.

For those who *do* believe, however, He provides everything needed in this present time to make us ready to enjoy His presence and fellowship throughout eternity. Furthermore, He has even told us about all these provisions in His written revelation to man, the Holy Scriptures.

He has told us also in His Word that the *real* problem with this world is sin, which can be defined as anything that comes "short of the glory of God" (Romans 3:23). That all people are sinners and in the same condemnation is obvious, for "there is none righteous, no, not one" (Romans 3:10).

Because there is sin in human hearts, there is ugliness in human lives — corruption, lust, duplicity, covetousness, hatred, fighting. The root of all such particular sins, however (as well as their corporate expressions in profiteering, racism, racketeering, war, and other large-scale social evils) is simply the basic sin of rebellion against God and His will. And the fundamental intellectual rationale for this age-long state of rebellion against God is always, ultimately, the concept of evolution (or whatever may be the contemporary name for the philosophy which seeks to explain and control the world without God, worshiping and serving "the creature more than the Creator" — Romans 1:25).

God, however, because of His nature of love, has provided a wonderful means of deliverance from sin and its effects (the final such effect necessarily otherwise being eternal separation from God). He has revealed His nature and purposes not only in written form (the Scriptures), but in living form, becoming man Himself!

In the person of Jesus Christ, God has demonstrated to all men His true purpose for man. In his humanity, Christ was the *perfect* man — man as God intended man to be. His death on the cross, therefore, provided full payment of all our debts to an offended Creator. "For He hath made Him to be sin for us, who knew no sin, that we might be made the righteousness of God in Him" (II Corinthians 5:21).

Forgiveness and reconciliation to God, on the basis of the substitutionary death and victorious resurrection of Christ, can now be received individually by anyone willing to repent (that is, literally, "change his mind") about God, and toward God, and to believe that Christ is his Creator, Judge and Saviour. Such a decision will change his life, change his world, and change his destiny.

For that person at least, the troubled waters of naturalistic human philosophy will be purged by the pure, living water of God's Word — "a well of water springing up into everlasting life" (John 4:14). The light of God's revelation will dissolve forever the long night of pagan evolution. "O taste and see that the Lord is good: blessed is the man that trusteth in Him" (Psalm 34:8).

Abiogenesis, *See* Spontaneous generation

Acquired characters, inheritance of, 55, 57, 61

Adaptation, 155
  See also Variations

Age
  of earth, 20-21, 68-69
  of man, 150-154

Allegorical interpretation, 10, 63-64

American Scientific Affiliation, 13, 14

Angels, 71, 74, 75

Animism, 38, 70

Anthropology, 28

Apologetics, 60

Artificial selection, 62

Astrology, 38, 72, 73-74

Astronauts, inter-planetary, 169

Astronomy, 29, 134-135

Atheism, 26-27, 34, 189.
  See also Humanism

Atomism, 64, 66-68

Babylon, source of pagan religions, 69-72, 73

Behavior, Human, 141, 177-178, 186

Bible
  banned from schools, 12
  contrary to evolution, 184-186, 188
  contrasted to geological-age system, 186-190
  creation model, 101-102, 107-109, 187
  divinely-inspired, 102, 185

Bible-Science Association, 15.

Biology, influence of evolution, 27-29, 33

B.S.C.S. textbooks, 11, 12

Buddhism, 38, 149

California Board of Education, 15

Canopy theory, *see* Waters above the Firmament

Carrying capacity of planet, 146, 184. *See also* Population

Cataclysm
  Interpretation of geologic formations, 93-96
  Noahic flood, 21-22
  Pre-Adamic, 10, 189-190
  See also Catastrophism, Flood

Catastrophism
  Biblical flood, 21-22, 108, 190
  Denied by Lyell and Darwin, 55-56
  Indicated in geologic formations, 21-22, 93-96
  Postulated by creation model, 106-107
  Primeval nucleogenesis, 132

Cause-and-effect, law of
  Contradicted by evolution, 31, 170
  Evidence of God, 104-105
  Taught by Aristotle, 65

Chemical predestination, 130

Chinese philosophy, 68
Christ
  Creator, 183
  Incarnation, 64
  Personal Saviour, 25, 141
  Redeemer, 160-161, 183-184
  Second coming, 142, 147,
    160
  Teachings inconsistent
    with evolution, 36
Christian Heritage College,
  15
Christianity, harmful influ-
    ence of evolution on,
    39-40, 63-64
Christian schools, 40, 178-181
Chronology, 20-22, 28, 68-69,
  150-154
Citizens' Movements for
  Creation, 15-16, 32
Classical thermodynamics,
  116-118
Classification of organisms,
  83-85
Coal, 158-159
Colleges, Christian
  Consortium, 179
  Evolutionist influence,
    178-181
Communism
  Distorted view of history,
    32
  Evolutionary basis, 41-42,
    186
  Racist doctrines, 45
  Revolutionary movements
    contemporaneous with
    Darwin, 59, 60

System of philosophy, 33
Comparative morphology
    and embryology, See
    Similarities and
    Differences

Complexity, See Entropy,
    Information, Order
Confucianism, 38, 149
Conservation
  Biblical references, 107,
    116
  Energy, 115-116, 158
  Entropy, 158-159
  Matter, 115
  Momentum, 115
  Principle of, 98, 106, 107
Consortium, Christian Col-
    lege, 179
Controlled evolution, 46-48
Conversion of energy, 100-
    101, 114-115, 126
Cosmogonies, evolutionary
    29, 65-72, 132-135
  Ancient religions, 65-72
  Modern theories, 29,
    132-135
Cosmological principle,
    132-133

Cosmology, See Universe
Creation model
  Contrasted with evolution
    model, 81-83, 105
  Definition, 81-83, 106-107
  Predictions confirmed
    Chronology, 153-154
    Laws of Thermodyna-
    mics, 97-98

Nature of biological
changes, 87
Physical characteristics
of rocks, 93
Systematic gaps in
fossil record, 88-90
Creation Research Society,
14-15, 96, 180
Creationism
Biblical revelation, 69, 101-
102, 107-109, 158-160, 183-
184
Ecological stewardship,
157-158, 160
Foundation of all truth,
183-184
Implementation in schools,
173-178
Legal implications, 172-173,
175
Modern Christian indiffer-
ence, 25-26
Modern revival by scien-
tists, 13-15, 182
Not capable of scientific
proof or disproof, 80
Creationist conferences, 18
Creator, See God, Creation-
ism
Crystals, entropic defects
in, 134
Cults, pseudo-Christian, 39
Culture, development of,
31-32, 33, 152, 170-171
Curse, Edenic, 107-108, 109,
142, 158-160, 188
See also Death, Disin-
tegration, Entropy,
Second Law of Thermo-
dynamics
Cybernetics, 119, 131

Darwin, life of, 51-58
Darwinian Centennial, 12,
37, 52
Darwinism
Acceptance of, 26
History of, 51-62
Social, 32, 41-42, 149, 156-
157
See also Evolution
Dating methods, 20-22, 28,
150-154
Day-age theory, 10, 180, 187-
188
Death
Fossils, 20, 94, 140, 159
Organisms, 105, 188, 190
Species, 139
Universe, 18, 117
Debates, creation-evolution,
16, 121
Deism, 59, 60
Deluge
See Catastrophism, Flood
Demonic activity, 38-39, 65, 74
Design in nature, 27, 61, 131
See also Order
Disease, 140
Disintegration
Cultures, 140-141
Environment, 140
Matter, 134; 140
Morals and religion, 141
Principle of, 17, 98-101, 106,
108, 112

Species, 139-140
See also Entropy, Second Law of Thermodynamics
DNA, 18, 125, 135, 137
Documentary theory of Old Testament, 39
Drosophila studies, 16, 80, 87

Ecology
Crisis, caused by evolutionary thinking, 156-157
Evolutionary relationships, 155
Initial perfection, 157-158
Relationships between organisms, 154-155, 157-158
See also Environment
Education, influence of evolutionism on, 32, 33, 35, 171-181
Energy
Conservation, 115-116, 158
Conversion, 100-101, 114-115 126
Creation, 141-142, 158
Crisis, 154
Definition, 114
Decreasing availability, 17, 108, 116-118
Future sources, 159-160
Initial abundance, 158-159
Required for growth, 19-20, 100-101, 122-123
Solar supply, 19, 100, 122, 123, 126-129, 158, 160
Entropy
Biblical references, 112

Conservation in initial creation, 158-159
Criteria for decreasing, 123-129, 133-134
Definition, 17, 112, 117
Efforts to harmonize with evolution, 122-123, 129-132
Increase of, 98-101, 111, 113, 118-119
Information systems, 119-121

See also Disintegration, Second Law of Thermodynamics
Enuma Elish, 70-71
Environment
Decay of, 140, 146, 149, 156
Deterioration blamed on Biblical mandate, 148-149, 155
Future renewal, 160-161
Initial perfection, 157-158
Pollution a result of evolutionary thinking, 156-157
Enzymes, 135
Epicureanism, 64, 66
Ethics
Christian, 177-178
Evolutionary, 34-37, 183, 186
Evaporites, 94
Evolution model
Contrasted with creation model, 82-83
Definition, 81-82, 112, 113
Failure of predictions
Continuum of organisms, 83-85

Development of elements and the cosmos, 29, 132-135

Human chronology, 150-154

Increasing complexity resulting from variations, 16, 85-88, 136-139, 163

Life from non-life, 135-136, 170

Physical changes in rocks with time, 93-94

Principle of increasing order, 17-19, 97-101, 111-113, 132

Transitional forms in fossils, 89-91

Not capable of proof or disproof, 80, 172

Supposed evidences of evolution, 56, 62, 82

Evolution Protest Movement, 13

Evolutionary processes

Future control, 46-48

Inefficient and cruel character, 19, 36, 43-44, 47, 156

Inheritance of acquired characters, 55, 57, 61

Mechanisms, 16, 85-88

See also Mutations, Natural selection

Rapid changes, 90, 91

Evolutionism

Anti-Biblical, 184-186

Anti-Christian, 36-37, 183, 185-186

Anti-theistic, 26-27, 34, 72-76, 189

As a complete cosmology, 26-27, 185

Behavior, influence on, 12, 34-37, 166-168, 186

Christian colleges, teaching in, 178-181

Christian periodicals, influence on, 181

Creationism, emotional prejudice against, 172, 176, 177

Exploitation of resources, responsible for, 149, 150, 156-157

History of, 51-76

Humanism, equivalent to, 33, 171-172

Legal aspects of teaching, 172-173, 175

Marriage, breakdown of, 166-168

Origin of, 72-76

Pre-Christian, 65-72

Pre-Darwinian, 55-65

Religious character of, 12, 22, 172

Sexual revolution, effect on, 166-168

Unscientific character of, 17-19, 80, 111-113, 132, 172

Evolutionist influence on human thought and teaching

Anthropology, 28

Biology, 27-29, 33

Education theory and methodology, 32, 33, 35, 171-181

Ethics, 34-37, 183-186

Geology, 20, 28

History, 32-33

Humanities and fine arts, 33-34

Mathematics, 29-30

Philosophy, 33

Physics and astronomy, 29, 132-135, 169

Psychology, 30-31, 36, 168

Racism, 41, 42, 44-46, 161-165, 186

Religion, 37-40, 83

Sociology, 31-32, 46

Totalitarian ideologies, 40-44, 186

Extinction of species, 139

Extra-terrestrial life, 168-171

Fascism, *See* Nazism

Fine arts, evolutionary influence on, 33-34

Flood, Biblical

Cataclysmic nature, 108, 109

Date of, 153-154

Geologic effects, 21-22, 60, 190

Traditions of, 22

See also Catastrophism

First Law of Thermodynamics, 98, 106-108, 114-116, 127, 158

See also Conservation

Fossil fuels, 158-159

Fossils

Contradictions in assumed evolutionary sequences, 21-22

Death, evidence of, 20, 94, 140, 159

Evolution, chief evidence for, 20, 28, 108, 189

Flood, probably formed by, 190

Gaps in record, 20, 88-92

Order of deposition, 92-93, 96

Purpose, lack of, 132, 156

Unconformities, 95-96

Friction, 117, 139

Fruit-flies, mutations on, 16, 80, 87

Gaps in fossil record, 20, 88-92

Gap theory, 10, 180, 182, 188-190

Genetic code, 18, 84, 125-126, 135

Genetic load, 87

Geologic column, 94-96

Geological ages

Contradictions with Bible, 180-181, 184-190

Contradictions with geologic evidence, 21-22, 93-95

Dated by stage-of-evolution, 189

Identified by fossils, 20, 89, 189

Supposed equivalence with days of creation, 10, 180, 187-188

Supposed insertion in gap in Genesis, 10, 180, 182, 188-190

Geology, evolutionary teaching in, 20, 28

Gnosticism, 62-63, 64

God
Character of, 105-106, 113, 132, 190
Existence of, 104-105
Inconsistent with evolution, 131-132, 185-186
Purpose of, 105, 131, 147, 148, 160-161
Work of, 107-108

Gospel of Christ, 184

Higher criticism, 39-40, 59

Hinduism, 38, 68, 149

History of evolutionary thought, 51-76

History, teaching of, 32-33

Humanism, 33, 35. *See also* Atheism, Pantheism

Humanities, evolutionary influence in, 33-34

Igneous rocks, 94

Illuminism, 59

Information systems
Creation of, 141
Genetic, 18-19, 84, 135-136
Thermodynamic, 119-121, 125-126, 130
See also Order, Program

Inspiration of Bible, 101, 185

Institute for Creation Research, 15, 96, 180

Islam, 39

Judaism, 39

Karma, 38

Kinds of organisms
Classification, 83-85
Extinction, 139
Origin, 136-139
Permanence, 98, 185

Languages, *See* Tongues

Laws on teaching of origins
Anti-evolution, 9
Permitting creation, 172-173, 175
Requiring creation, 15-16

Life
Complex organization, 100
Extra-terrestrial, 168-171
Origin, 135-136, 170

Life-Time books on evolution, 12

Limestone, origin of, 94

Lithology, 21-22, 94

Man
Created as God's steward, 157
Future control, 46-48, 145-149
Origin of present population, 150-154
Technological skill in antiquity, 170-171

Marriage, influence of evolutionary teaching, 166-168
Mathematics
Influence of evolutionary thinking, 29-30
Population growth models, 150-153
Matter
Conservation of, 115
Origin of, 132-135
Membrane, 128
Milesian philosophy, 66-68
Missionaries, 38
Momentum conservation, 115
Monotheism, 38, 72
Moody Institute of Science, 13, 14
Morality, 141, 186
Moses, 39, 70
Motion, 139
Mutations
Consistent with entropy principle, 19, 138
Drosophila studies, 16, 80
Evolutionary mechanism, 16, 86-87, 136-138
Genetic "mistakes," 18
Harmful, 87-88, 137-138
Random nature, 100, 126, 128, 136-137, 156
See also Natural selection
Mystery religions, 64-65, 69-70, 73
Mysticism, 59, 63

Nascent organs, 140

N.A.S.A., 169
National Science Foundation, 12
Nations
Decay of, 140-141
Origin of, 108, 162, 165
Natural selection
Artificial selection, 62
Competition in human societies, 37
Conservational process, 19-20, 86-87, 126, 129, 139
Darwin's claim to theory, 55-57
Evolutionary mechanism, 16, 62, 86-87, 100-101, 138-139, 156
Influence on cultural developments, 33
Origin of theory, 52-53, 54-55
See also Mutations
Nazism, 32, 33, 40-41, 43, 45, 165, 186
Negro, 164-165
Nimrod, 71, 72, 73, 74
Noahic prophecy, 162
Nucleogenesis, 132-133

Occultism, 38-39, 65.
Oil, 158-159
Open systems
Effect on geologic dating, 21
Required for growth in order, 18-19, 99-100, 122-123, 124

Order of fossils, 21-22, 92-93, 96
Order, degree of
  Creation of, 141-142, 158
  Decrease required by entropy principle, 113, 118-119, 130
  Living organisms, 100, 134-135
  Requirements for increase, 18-19, 99-101
  Supposed product of evolution, 97
Organisms, living
  Classification, 83-85
  Complexity, 100, 134-135
  Ecological relationships, 154-155, 157-158
  Kinds, 136-139, 163
  Mechanisms of variation, 16, 85-88
Oscillating Universe theory, 133.

Paganism, 62-63, 64-65, 65-72
Pantheism
  Advocated by environmentalists, 149-150
  Equivalent to evolutionary humanism, 33, 35
  Equivalent to polytheistic paganism, 38, 39, 64-65
  Mysticism, 59, 63
  Pre-Christian philosophies, 66, 69, 72
Peppered moth, 16, 80, 86, 87

Periodicals, Christian, 181
Perpetual motion, 139
Philosophy
  Evolutionary, 33. *See also* Evolutionism
  Pantheistic, 33, 35, 38, 39, 59, 64-65
  Pre-Christian, 64-72
Photosynthesis, 19, 100, 127
Physics, evolutionist teaching in, 29, 132-135, 169
Pollution, 156, 157, 159-160
  See also Environment
Polytheism, 65, 69-72, 73, 149
  See also Paganism, Pantheism.
Population, human
  Future growth, 47, 145-149
  Limitation of, 145, 149
  Malthus theories, 55, 150
  Recent origin, 150-154
Power, 114, 141-142
  See also Energy.
Predestination, chemical, 130
Program
  Required for growth in order, 100, 124-126, 130, 139
  Requires initial programmer, 19
  See also Information, Order
Progressive creation, 92, 179, 180, 186
  See also **Theistic evolution**

Psychology, evolutionary teaching in, 30-31, 36, 168

Public schools, 11-12, 22, 32, 171-178

Purpose in history
Implied in character of God, 105-106, 113, 131-132, 147, 148, 160-161, 190
Not evident in fossils, 132, 156

Races, origin of, 161-163

Racism
Based on evolution, 41, 42, 44-46, 162-165, 186
Espoused by 19th Century evolutionists, 45, 162, 165-166
Not found in Bible, 44, 45, 161-162, 165-166

Radiations
Mutations caused by, 137
Nuclear testing, 156
Solar energy, 19, 100, 122, 123, 129, 158, 160

Radiometric dating, 20-21, 196

Reformation Period, 59, 63 64

Religion
Decay of, 141
Evolution a religious system, 12, 22, 171-172
Influence of evolutionary thought, 37-40, 83
Unity of all ancient systems, 69-72

Revolutionary movements, 41-44, 59-60
See also Communism, Nazism

Rocks, physical characteristics of, 21-22, 93-95

Salvation, 25, 141-142, 160-161, 183-184

Satan, 71, 74, 75-76, 190

Schools, evolutionary teaching in
Christian schools, 40, 178-181
Citizens' movements against, 15-16
Public schools, 11-12, 22, 32, 171-178

Scientific method, 80, 172

Scientists, Bible-believing 13-15, 60-61

Scopes trial, 9, 11, 25, 175, 182

Second Law of Thermodynamics
Apparent exceptions, 17-18, 97-101, 121-123, 138
Criteria for apparent exceptions, 123-129, 133
Decreasing availability of energy, 116-118
Definition, 17, 98
Equivalence of different forms, 120-121
Evolution, attempted harmonizations with, 122-123, 129-132

Increasing disorder, 118-119
Information theory, 119-121
See also Disintegration, Entropy
Sedimentary processes, 94-95
Sermons from Science, 14
Sexual revolution, influence of evolutionary teaching, 116-168
Shintoism, 38
Similarities and differences in organisms, 83-85
Sin, cause of the Curse, 107-108, 188, 190
Social Darwinism, 41-42, 149, 156-157
Socialism, 59-60, 165
Social sciences, influence of evolutionism, 32
Society for Study of Creation and Deluge, 13
Sociology, 31-32, 46
Solar system, origin of, 134-135
Soul-winning, excuse for ignoring evolution issue, 9, 176, 180-181
Special creation, 104, 106, 107, 109
See also Creation model
Spiritism, 38, 72
Spontaneous generation, 64, 65, 66, 104, 170
Stars, evolution of, 134-135
Statistical thermodynamics, 118-119, 122, 129-130

Stoicism, 64, 65
Stratigraphy, 92-93, 94-95, 96
Struggle for existence, 41, 42, 57, 59, 149, 157, 163, 164, 185
Student attitudes, teaching of evolution, 176-178
Suffering, problem of, 106, 188, 190
Sun, as source of energy for earth processes, 19, 100, 122, 123, 126-129, 158, 160
Survival of the fittest, 41, 42, 57, 59, 80, 156, 157, 163, 185

Taoism, 38
Taxonomy, 83-85, 92
Teleology
    Evidence of design in nature, 61
    Evolutionists, repudiated by, 131
    Milesian philosophy, early rejection by, 66-68
    Natural selection, replaced by, 27
Textbooks, 11-12, 15, 29, 32, 174, 175
Theism, *See* Creation model, God
Theistic evolution
    Attempt to harmonize entropy with evolution, 131-132
    Christian colleges, teaching in, 131-132
    Contrary to Bible, 63, 185-186

Not a scientific model, 92-93, 102-103
Rejected by evolutionary leaders, 102, 103, 104-105 131-132
Unsatisfactory compromise, 10-11, 43
Thermodynamics
Classical, 116-118
Discipline of, 113-114
First Law, 98, 108, 114-116, 127, 158
Information theory, 119-121
Laws of, 98, 108, 113
Statistical, 118-119, 122, 129-130
Second Law, 17-18, 98,101, 108, 116-118
Testimony to creation, 117-118, 141-142
See also Energy
Time's arrow, 17
Tongues
Confusion of, 108-109
Decay of, 140-141
Divisions among men, 161-162, 165
Totalitarianism, 40-44, 186
Tower of Babel, 74, 108
Traditions of Noahic flood, 22

Unconformities, 95-96
Unidentified flying objects, 168-169
Uniformitarianism
Assumed in geologic dating methods, 21
Basic in geologic interpretation, 93
Foundation of Darwinism, 55-56, 62
Opposed to catastrophism, 21-22, 25-26
Universe, origin of, 29, 132-135

Variations, 16, 85-88, 136-138
See also Mutations
Vestigial organs, 140
Victorious living, 9, 182

War, 36, 41, 43, 47
Water
Above the firmament, 64, 158
Primordial matter, 71
Wear, 140
Work, 114, 117
World Population Year, 146

Zodiac, 73, 74

INDEX OF NAMES

Anaximander, 66
Anaximenes, 66
Aquinas, Thomas, 63
Aristotle, 63, 64, 65, 66
Asimov, Isaac, 99, 115, 116, 119, 122, 133
Augustine, 63
Ayala, Francisco J., 103, 121, 131, 132, 137

Ball, John, 170
Barlow, Nora, 57
Barnes, Harry Elmer, 31
Barzun, Jacques, 42, 43, 53, 54
Basil, 63
Beard, Charles, 33
Beck, W. S., 100, 125
Beesley, E. M., 93
Berelson, Bernard, 146
Berkowitz, Norman, 32
Berry, W. B. N., 28
Birch, L. C., 155
Blinderman, Charles S., 58
Blyth, Edward, 55
Blum, Harold, 118
Brain, Walter R., 34
Brown, Arthur I., 13
Bryan, William Jennings, 9, 10
Burhoe, Ralph, 47
Burnham, John C., 45
Butler, Joseph, 60

Cavalli-Sforza, L. L., 163
Chambers, Robert, 55
Coale, Ansley J., 151
Conford, F. M., 67

Coon, Carleton, 165
Crick, F. H. C., 170

Darlington, C. D., 28, 55, 56, 57, 59
Darrow, Clarence, 9, 10
Darwin, Charles, 12, 33, 36, 41, 42, 45, 51, 52, 53, 54, 55 56, 57, 58, 59, 61, 91, 103, 135, 150, 162, 164, 165
Darwin, Erasmus, 55, 56
Davis, D. Dwight, 90, 91
Democritus, 64, 66
Descartes, Rene, 65
Dewey, John, 32, 33, 35
Diderot, Denis, 55
Dingle, Herbert, 133
Dobzhansky, Theodosius, 27, 43, 81, 82, 84, 85, 97, 138, 171, 172
Dreyer, J. L. E., 64
Dubos, Rene, 30, 34, 81, 82
Dwight, Timothy, 60
Dyson, Freeman J., 120

Eddington, Arthur, 17
Edwards, Jonathan, 60
Ehrlich, Paul R., 155
Einstein, Albert, 130
Eiseley, Loren C., 53
Engels, Friedrich, 42

Faraday, Michael, 60
Franklin, Benjamin, 55
Freedman, Donald, 146
Freud, Sigmund, 30, 168

George, T. N., 91
Grabiner, Judith V., 11, 12
Graham, W., 164
Gregory of Nyssa, 63

Haeckel, Ernst, 62
Haller, John S., 44, 165
Haskins, Caryl P., 135
Hegel, G. W. F., 59
Heidel, Alexander, 70
Hesiod, 68
Himmelfarb, Gertrude, 41, 43, 164
Hislop, Alexander, 69, 70
Hitler, Adolf, 41, 43, 45, 186
Hoagland, Hudson, 31, 47, 48
Hulse, Frederick S., 88
Huxley, Julian, 12, 13, 37, 81, 82, 103, 113, 137, 138
Huxley, Leonard, 105
Huxley, Thomas, 10, 35, 58, 61, 105, 164, 165

Jacobsen, Thorkild, 71
Jacobson, Homer, 136
James, Henry, 30
Jepsen, Glenn, 91
Joule, James P., 114
Jung, Carl, 30

Kant, Immanuel, 65
Keith, Arthur, 36, 37, 43, 46
Kelvin, William Thomson, 60, 114

Lamarck, Jean B., 42, 55, 56, 57, 61

LaPlace, Pierre Simon, 65
Latham, R. E., 67
Laurence, William, 55
Leibnitz, Gottfried, 58
Leucippus, 64, 66
Levine, R. P., 127
Lewin, Roger, 170
Lindsay, R. B., 113, 116
Lindsey, Arthur Ward, 63
Lomax, Alan, 32
Lucretius, 64, 67
Lyell, Charles, 52, 55, 56, 61
Lyttelton, Lord, 60

Malthus, Thomas, 52, 55, 150
Martin, C. P., 88
Marx, Karl, 33, 42, 45, 54, 59, 60, 61, 186
Maxwell, William Clerk, 60, 114
Mayr, Ernst, 26, 27, 52, 91
McIrvine, Edward C., 120
Mead, Margaret, 146, 148
Means, Richard L., 149
Mendel, Gregor, 16
Miller, Peter D., 11, 12
Mintz, Sidney M., 44
Moody, Paul A., 28, 90
Moore, John N., 14, 29
Morris, Henry M., 96, 102, 152
Motulsky, Arno G., 34, 48, 183, 184
Muller, H. J., 33, 35, 47
Munitz, Milton K., 64, 66, 67, 70, 71
Mussolini, Benito, 41, 43

Newman, J. R., 113
Newton, Isaac, 58, 60, 114
Nietzsche, Friedrich, 33, 41, 59, 60, 165

Ogilvey, C. S., 30
Orgel, Leslie, 170
Origen, 63

Paley, William, 60
Pascal, Blaise, 60
Plato, 65, 66
Plochmann, George K., 58
Popper, Karl, 80
Powicke, F. M., 40
Price, George McCready, 13
Pritchard, James C., 55
Ptolemy, 63

Rabinowicz, Ernest, 140
Revelle, Roger, 148
Rimmer, Harry, 13
Rogers, Carl, 168
Romanes, George J., 36

Scopes, John T., 9, 10, 11, 25, 175, 182
Seitz, Frederick, 37
Sigurbjornsson, Bjorn, 137
Simar, Mitton, 170
Simpson, George Gaylord, 46, 56, 57, 91, 99, 100, 125, 162
Skinner, B. F., 168
Slusher, Harold S., 29
Smith, Albert J., 179
Socrates, 65

Spencer, Herbert, 57, 58, 62
Spinoza, Baruch, 65
Stanton, Mary, 32

Tax, Sol, 68, 103
Thales, 66, 67, 68, 70
Thomas, Lewis, 128
Tobey, James A., 140
Toynbee, Arnold, 33
Tribus, Myron, 120

Ussher, James, 60, 153, 154

Veith, Ilza, 68

Waddington, C. H., 36
Wagner, Wilhelm Richard, 54
Wald, George, 104
Wallace, Alfred R., 52, 53, 46
Watson, D. M. S., 104
Watson, John B., 30, 168
Weber, Christian O., 33
Weisskopf, Victor K., 29
Wells, William R., 55
West, Gilbert, 60
Wheeler, H. E., 95
Whitcomb, John C., 96
White, Lynn, 149
Wick, Gerald L., 169
Wiedmann, Jost, 96
Woodward, John, 60

Zirkle, Conway, 42, 45

INDEX OF SCRIPTURES

GENESIS

1:1	184,189
1:2	10,184,189,190
1:5	187
1:7	158
1:11	158
1:16	190
1:17	160
1:20	158
1:24	158
1:26-28	157,166
1:28	147,149
1:29	157
1:30	157
1:31	157,158,190
2:1-3	158
2:2	190
2:3	107,116
2:15	157
3:17	107,158
6:9	190
9:1	147
9:25-27	162
10	73
10:5	161
10:8-10	71
10:20	161
10:31	161
11	73,154
11:4	74
11:6,9	161

EXODUS

20:11	187

I KINGS

18:21	181

NEHEMIAH

9:6	158

JOB

38:4-7	157

PSALMS

19:1	157
19:6	160
34:8	194
35:26	112
44:15	112
69:19	112
71:13	112
102:26	158
104:24	161
104:31	161
109:29	112
127:3-5	147
128:1-6	147
148	157
148:6	158

PROVERBS

15:1	177
16:7	178
25:11	177

ECCLESIASTES

3:14	116

ISAIAH

14:12-14	74
14:12-15	190
14:14	76
40:26-31	141
40:31	142

JEREMIAH
2:27   112
51:7   70

EZEKIEL
28:11-17   190
28:12,13   75
28:17   75,76

MATTHEW
7:18,19   168
7:18   186
19:3-9   167
19:4-6   185

JOHN
4:14   194

ACTS
17:26   162,166
17:27   166

ROMANS
1:18-32   38
1:21-25   112
1:23   73
1:25   73,193
1:28   73
3:10   193
3:23   193
5:12   188,190
7:21-25   141
8:20   142,158
8:20-22   160
8:21   107,142
8:22   108,142,158,188

I CORINTHIANS
6:5   112
10:31   178
14:33   188
15:21   190
15:34   112

II CORINTHIANS
4:4   75,126
5:18   161
5:21   193

GALATIANS
4:1,2   177

EPHESIANS
1:14   160
5:3-5   167
6:3   147
6:5-8   177
6:12   74

COLOSSIANS
1:16   183
1:20   142,160,183
3:22-24   177
3:23   178
4:6   177

I TIMOTHY
3:6   75

II TIMOTHY
2:15   178
2:24,25   176

TITUS
  2:9,10   177

HEBREWS
  1:3   107,116,158
  13:4   167

JAMES
  1:17   112

I PETER
  3:15   178

II PETER
  3:4   112
  3:7   107
  3:16   108

REVELATION
  3:14   183
  4:11   157
  5:13   157
  7:9   161
  12:4   74
  12:9   75
  14:6,7   184
  17:5   72
  21:5   142,160
  22:3   160

FOR FURTHER READING

For those who are enough concerned about the evolutionary problem to want to do something about it in their own schools and communities, the first essential is to become sufficiently informed so that they can present the scientific case for creation intelligently and persuasively. The following books will be found especially helpful in this regard. Although not meant as a complete bibliography, the list below does cover all important aspects of these and related subjects. Those noted as technical monographs require scientific and mathematical backgrounds for full understanding; the others can be understood by any serious lay reader. They can be obtained at your local bookstore or ordered direct from CLP Publishers, P. O. Box 15666, San Diego, California 92115. Write for a complete descriptive catalog. Video catalog also available upon request.

What is Creation Science?
Henry M. Morris, Ph.D. and Gary E. Parker, Ed.D.
The question everyone is asking answered by outstanding educators and scientists in language you don't have to be a scientist to understand. Nearly 60 illustrations, with comprehensive indexes and bibliography. **No. 187**

Evolution? The Fossils Say NO! *Duane T. Gish, Ph.D.*
The most extensive treatment in print showing the universal and systematic gaps in the supposed fossil record of evolutionary history, including a thorough discussion of human origins, showing conclusively that man did not evolve from ape-like ancestors. Generously illustrated. Over 100,000 in print.
General Edition, No. 054; Public School Edition, No. 055

Evolution in Turmoil *Henry M. Morris, Ph.D.*
Sequel to *The Troubled Waters of Evolution*. Updating the status of evolutionary thought and actions in recent years. **No. 271**

Men of Science/Men of God *Henry M. Morris, Ph.D.*
One of the most serious fallacies of modern thought is that genuine scientists cannot believe the Bible. Illustrated, brief biographies of some major scientists who believed they were "thinking God's thoughts after Him." **No. 108**

Creation: The Facts of Life *Gary E. Parker, Ed.D.*
If you are an "armchair scientist" with a hunger for knowledge (but no Ph.D. to help you understand it!), this book was written for you. Clear explanations and "down home" examples of the basic concepts of creation and evolution **No. 038**

The King of Creation *Henry M. Morris, Ph.D.*
Places modern creation movement in its biblical perspective, emphasizing Christ as Creator and Sovereign of the world. **No. 096**

Acts & Facts/Impact Series Anthology
The past decade has seen a great increase in interest and controversy in the realm of creationism. These five books trace this growth through the compilation of significant articles and debates that have been reported in ICR's popular *Acts & Facts* publication. Although they may be ordered individually, the complete five-volume set provides a comprehensive record of the recent revival of interest in creation science.

Creation: Acts/Facts/Impacts, 1972-1973	**No. 037**
The Battle for Creation, 1974-1975	**No. 013**
Up With Creation, 1976-1977	**No. 179**
Decade of Creation, 1978-1979	**No. 044**
Creation—The Cutting Edge, 1980-1981	**No. 272**

Scientific Creationism *Henry M. Morris, Ph.D., Ed.*
The most comprehensive and up-to-date textbook or reference handbook now available covering all major aspects of the field of scientific creationism. Available also in a special Public School Edition in which all biblical references and discussions are omitted.
General Edition No. 140, Paper
Public School Edition No. 141, Paper; No. 357, Cloth

Many Infallible Proofs *Henry M. Morris, Ph.D.*
A complete reference handbook on all aspects of practical Christian evidences, for the strengthening of personal faith in the inspiration of the Bible and the truth of Christianity. Believed to be the most comprehensive and up-to-date textbook available on this subject. Evidence from science, history, prophecy, internal structure, philosophy, and common sense, with answers to the various alleged mistakes and contradictions of the Bible.
Cloth, No. 103; Paper, No. 102

The Bible Has The Answer (Expanded Edition)
Henry M. Morris, Ph.D. and Martin E. Clark, D.Ed.
Scientific, logical, and biblical answers to 150 frequent questions on the Bible and science, occultism, controversial doctrines, person and work of Christ, the practical Christian life, modern world problems, things to come, and many others. Baptistic and premillennial on doctrinal questions. **No. 023**

The Genesis Flood *John C. Whitcomb, Th.D. and Henry M. Morris, Ph.D.*
The standard definitive text in the field of scientific biblical creationism and catastrophism; the most extensive and best-documented treatment available on both the biblical and scientific implications of creation and the flood. Largely responsible for the present revival of scientific creationism. **No. 069**

The World That Perished *John C. Whitcomb, Th.D.*
A sequel to *The Genesis Flood,* with refutations of criticisms and further evidences of a young earth and biblical catastrophism. Strikingly illustrated. **No. 184**

The Genesis Record *Henry M. Morris, Ph.D.*
Verse by verse scientific and devotional commentary on the book of beginnings. **No. 070, Cloth**

The Early Earth *John C. Whitcomb, Th.D.*
Studies of the origin and nature of man, the gap theory, the antediluvian world, and others. Illustrated. Dr. Whitcomb is Professor of Old Testament and Director of Post-Graduate Studies at Grace Theological Seminary. **No. 051**

The Bible and Modern Science *Henry M. Morris, Ph.D.*
An evangelistic presentation of evidences for the scientific validity of the Bible, including archaeological and prophetic confirmations. Over 300,000 in print. **No. 333**

Dinosaurs: Those Terrible Lizards *Duane T. Gish, Ph.D.*
Did dinosaurs live at the same time that humans did? Are dragons just imaginary creatures? In this beautifully illustrated book for children, Dr. Gish explains what dinosaurs were and why they no longer exist. The issue of creation vs. evolution is presented on a level that children can easily understand. **No. 046, Cloth**

Dry Bones . . . and Other Fossils *Gary E. Parker, M.S., Ed.D.*
What are fossils? How are they formed? What can we learn from them? These and many other questions are answered in conversational dialogue in this creatively illustrated book for children. Its educational value is enhanced by including references to the fall of Adam, the Flood, and the promise of a new earth. An explanation of fossils and their significance presented in a manner that children will understand and enjoy. **No. 047**

Tracking Those Incredible Dinosaurs *John D. Morris, Ph.D.*
What's the *real* story on those footprints in the Paluxy River bed? What do they really tell us? An eye-witness report documented by nearly 200 photos. Dr. Morris, highly qualified in geoscience, examines the historical evidence, as well as the existing evidence found in a nearby river of the small Texas town of Glen Rose, which has recently become the center of much controversy. Did man and dinosaurs live together in ancient times? *Look at these photos and draw your own conclusions.* **No. 173**

The Natural Sciences Know Nothing of Evolution
A. E. Wilder-Smith, Ph.D.
Examines the evidence and presents the conclusions in a comprehensive analysis of evolution from the viewpoint of the Natural Sciences. **No. 110**

Biblical Cosmology and Modern Science *Henry M. Morris, Ph.D.*
Scientific and biblical expositions of many aspects of cosmology, covering origins, catastrophism, demography, sedimentology, thermodynamics and eschatolgy. Includes extensive critiques of day-age, gap, and allegorical theories of Genesis. **No. 337**

Studies in the Bible and Science *Henry M. Morris, Ph.D.*
Sixteen studies on special Bible-science topics, including evidence of Christ and the Trinity in nature, the Bible as a scientific textbook, biblical hydrology, concept of "power" in Scripture, scientism in historical geology, and others. **No. 377**

Institute for Creation Research Technical Monographs

No. 1 Speculations and Experiments Related to the Origin of Life (A Critique) *Duane T. Gish, Ph.D.*
An analysis and devastating critique of current theories and laboratory experiments which attempt to support a naturalistic development of life from nonliving chemicals **No. 158**

No. 2 Critique of Radiometric Dating *Harold S. Slusher, Ph.D.*
Sound principles of physics are used to evaluate and refute the most important radiometric methods of determining geologic ages.
No. 159

No. 4 Origin and Destiny of the Earth's Magnetic Field
Thomas G. Barnes, M.S., Sc.D.
A technical exposition of one of the most conclusive proofs that the earth is less than 10,000 years old. **No. 161**

No. 5 Our Amazing Circulatory System . . . By Chance or Creation? *M. E. Clark, M.S.*
A technical study of the human heart and circulatory system, emphasizing the impossibility of evolution. **No. 162**

No. 6 Age of the Solar System
Harold S. Slusher, Ph.D. and Stephen Duursma, M.S.
Provides detailed mathematical evidence that the "Poynting-Robertson effect" (fall of interplanetary dust into the sun as a result of solar radiation pressures) requires a very young solar system. **No. 163**

No. 7 Age of the Earth
Harold S. Slusher, Ph.D. and Thomas Gamwell, M.S.
This in-depth study of evidence from the earth's thermal cooling and radioactivity shows that the earth cannot be old. **No. 164**

No. 8 Origin of the Universe *Harold S. Slusher, Ph.D.*
Examination of the Big-Bang and Steady State Cosmogonies. Shows convincingly that the universe could not have originated by naturalistic processes and favors a recent origin. **No. 165**

No. 9 Age of the Cosmos *Harold S. Slusher, Ph.D.*
Setting all assumptions and guesswork aside, this study looks at the physical indicators to the upper limits of the age of the cosmos.
No. 166